Image and Sound Processing for Social Media

IMAGE AND SOUND PROCESSING FOR SOCIAL MEDIA

Sourabh Pal

ARCLER PRESS

www.arclerpress.com

Image and Sound Processing for Social Media

Sourabh Pal

Arcler Press
224 Shoreacres Road
Burlington, ON L7L 2H2
Canada
www.arclerpress.com
Email: orders@arclereducation.com

ISBN: 978-1-77469-449-7 (Hardcover)

ABOUT THE AUTHOR

Saurabh Pal received his M.Sc. in Computer Science in 1996 and obtained his Ph.D in 2002. He then joined the Department of Computer Applications, VBS Purvanchal University, Jaunpur as a Lecturer. Currently, he is working as Professor. He has authored more than 100 research papers in SCI/Scopus in international/national conference/journals as well as authored four books and also guided many research scholars in computer science/applications. He is an active member of CSI, Society of Statistics and Computer Applications and working as editor, member of editorial board for more than 15 international journals. His research interests include bioinformatics, machine learning, data mining, and artificial intelligence.

TABLE OF CONTENTS

LIST OF FIGURES

LIST OF TABLES

LIST OF ABBREVIATIONS

2D	two-dimensional
3D	three-dimensional
AAC	augmentative and alternative communication
AHP	analytic hierarchy processes
AI	artificial intelligence
ASR	automatic speech recognition
B/W	black and white
BRIR	binaural room impulse response
CASA	computational auditory scene analysis
CNN	convolutional neural network
DEM	digital elevation model
DN	digital number
DTS	digital theater system
EDM	emergency decision making
FCC	false color composites
HCI	human-computer interaction
HOG	histogram of oriented gradients
HRIR	head-related impulse response
HRTFs	head-related transfer functions
ICA	independent component analysis
IED	interaural envelope delay
ILD	interaural level difference
ILSVRC	ImageNet large scale visual recognition challenge
ITD	interaural time difference
J2EE	java enterprise edition
LDA	latent Dirichlet allocation
MEMS	microelectromechanical systems
MIDI	musical instrument digital interface

MIR	music information retrieval
MTB	motion-tracked binaural
NER	named entity recognizer
OLI	operational land imager
PCM	pulse code modulation
RF	random forest
ROC	receiver operating characteristic
SAC	spatial audio coding
SBTF	standby task force
SFCC	standardized false color composite
SIFT	scale invariant feature transform
SNR	signal-to-noise ratio
SVMs	support vector machines
TCC	true color composite
UN OCHA	United Nations Office for the Coordination of Humanitarian Affairs
WFS	wave-field synthesis
YOLO	you only look once

PREFACE

The primary goals of this book are to give an introduction to basic ideas and approaches for image and sound processing for social media, as well as to provide the groundwork for further study and research in this topic. To accomplish these goals, we concentrated on content that we feel is basic and whose use is not confined to the solution of particular issues. Sound is omnipresent and essential in the lives of billions of people worldwide. Sound works and performances are among the most complex and complicated of our cultural objects, and sound's emotional power may impact us in unexpected and profound ways. From basic, solo folk tunes to popular and jazz sound, to symphonies with entire orchestras, sound encompasses a vast spectrum of forms and styles. The digital revolution in sound delivery and storage has sparked enormous interest and attention in the ways that information technology might be used to this type of material. Computers are now extensively involved in practically every element of sound consumption, from exploring personal collections to discovering new artists to controlling and safeguarding the rights of sound composers, not to mention their crucial involvement in most of today's sound creation.

Despite its relevance, sound processing is still a new subject when compared to speech processing, a study topic with a lengthy history. In fact, in the year 2000, the International Society for Sound Information Retrieval (ISSIR) was created, which systematically deals with a wide variety of computer-based sound analysis, processing, and retrieval subjects. Traditionally, computer-based sound research has relied heavily on symbolic representations such as sound notation or MIDI representations.

Because of the growing availability of digital audio information and the expansion of computing power, automated processing of waveform-based audio signals is becoming a more prominent focus of research efforts. Many of these research activities are focused on the creation of technology that enable people to access and study sound in all of its forms. Audio fingerprinting techniques, for example, are increasingly integrated into commercial programs that assist users in organizing their own sound collections. Extended audio players employ sound processing techniques to highlight the current measures within sheet sound while playing back a symphonic recording. Additional information regarding melodic and harmonic progressions, as well as rhythm and tempo, is automatically supplied to the listener on demand. Interactive sound interfaces highlight structural components of the current piece of sound and allow users to navigate straight to any essential segment, such as the chorus section, the primary soundal theme, or a solo passage, without the need for tiresome fast-forwarding and rewinding.

There are eight chapters in the book. The first chapter provides readers with an overview of picture and sound processing foundations. The second chapter discusses picture processing by integrating human and machine computation. Chapter 3 delves deeply into the use of online social media for disaster management.

Chapter 4 delves deeper into comprehending and categorizing picture tweets. The fifth chapter focuses on the usage of digital picture processing. The synthesized sound properties and self-presentation are demonstrated in Chapter 6. Chapter 7 discusses the current and future of sound processing. Finally, Chapter 8 focuses on mobile phone processing in the future of social media.

—***Author***

CHAPTER 1

FUNDAMENTALS OF IMAGE AND SOUND PROCESSING

CONTENTS

1.1. INTRODUCTION

There is a wide variety of mediums and formats available for the representation of sound. As an example, a composer could record a piece by writing it down as part of a Sound score. In a scoring, Soundal signs are utilized to graphically express notations as well as how a Soundian is expected to perform these notations. Sheets Sounds is another name for the printed form of a Sound score, which is also known as a score. Paper was the initial medium for this depiction; however, it may now be seen on computer screens as digital images thanks to technological advancements (Deserno, 2010). Basic rules, like the widely utilized Musical Instrument Digital Interface (MIDI) protocol, allow for the communication of sound for electronics and computer systems. Within these protocols, event texts can specify pitches, velocity profiles, and other parameters to generate the desired sounds. MIDI is one of the most commonly used protocols. We just use word symbols to denote every machine-readable datatype that expressly depicts Soundal objects throughout this book. Such Soundal objects may be anything from time note occurrences, as in MIDI files, to graphical forms with associated Soundal meanings, as in Sounds engraved systems (Tohyama et al., 2000).

Music representations like WAV or MP3 files, unlike symbolic forms, need not specify Soundal occurrences. Acoustics waves are created when a source (e.g., an instrument) produces sounds that pass to an auditory system as pressurization vibrations. There are three types of audio representations: sheets sound, symbol sound, an auditory sound. To put it differently, the word sheets Sound refers to visualizations of a score that are print or computerized visuals (Mannan et al., 2000). Any type of scoring representation with just a specific coding of notes or even other Soundal occurrences are referred to be symbolism. Lastly, the word audio relates to acoustical acoustic waves forms. Each one of these representations represents different qualities of the a Soundal body, but none of them captures all of its characteristics. Under this respect, each depiction may be thought about as a projecting or manifestation of the what we call an audio piece. Using these various Sound representations, we address some fundamental aspects of Sounds in this introduction part (Schreiber, 2012). To begin, we'll go over the fundamentals of West Sound composition as it applies to sheets Sounds representations. Although precise Sounds notation criteria are not required for this book, we do want a fundamental understanding of the frequency, length, and arrival duration of Soundal tones. Now we go through the fundamental features of symbols, focusing on MIDI, that is the most widely accepted model for managing sound synthesizers. Using systems ranging from basic circuits to complex parallel processing, modern electronic technology has made it feasible to manage multi-dimensional information. This manipulation's purpose may be categorized into three groups (Müller, 2015):

- Image Processing input → output
- Image Analyzer image in → measurements out
- Image Recognition image in → high-level description out

We'll concentrate on the fundamentals of image processing. We can only offer a few initial words regarding image processing due to space constraints. Analyzing images requires an approach that is profoundly different from the book's content. We'll also stick to two-dimensional (2D) photo editing, although most of the ideas and methods we'll cover may simply be expanded into three or even more degrees is recommended for readers who want more information than what is offered here or who want to learn about other parts of image processing (Toriwaki and Yoshida, 2009). Lastly, we talk about auditory representations, that are central to a book. We are particularly interested in characteristics of sound wave qualities including frequency, velocity, and resonance. We'll start with some fundamental definitions (Smith and Finnoff, 2009). In the "real world," each image is defined as a function of two different variables, such as an (x, y) with just an as that of the image's loudness (e.g., brightness) just at true coordinates point (x, y). Posts, also known as areas, ROIs, or simply zones, may be found in a single image. This idea stems from the fact that photographs usually include groups of items, which each might serve as the foundation for an area. It should be feasible to apply certain image processing procedures to specified locations inside a comprehensive image processing and analysis. As a result, one section of an image (region) may be treated to reduce blurriness while the other is handled to enhance color rendering (Kuo et al., 2013).

A given image's amplitudes would nearly be always among real or integers values. The latter would be frequently the outcome of a quantized procedure that reduces a constant range (like 0 to 100%) to a fixed number of different levels. Therefore, in certain photosystems, the signals may incorporate photons counting, implying that now the intensity would be quantization intrinsically. This purely physical measurement yields a complicated amount in the type of an actual magnitude as well as a real period in those other image processes, including such magnetic resonance imaging. Except as otherwise stated, we shall treat frequency and amplitude like reals or integers for the rest of this book (Korpel, 1981).

This section includes the most important image processing techniques for remote sensing techniques, including image display, quantitative methods, and theme information retrieval. From fundamental visualization methods which can be used to enhance digital camera photographs to the more difficult multidimensional transform-based approaches, this sequence of sections presents subjects having increasing complexity (Schueler et al., 1984). A digital image may increase the image quality of photographs, dynamically improve and emphasize certain image aspects, and categorize, analyze, and recover spatial and spectral patterns that convey thematic data from images. This can adjust the image's geometric features and light levels at will to provide distinct viewpoints of the same image. Image editing, for example, cannot add information to actual visual data, but it may improve presentation so that we can notice more from improved photos than from the original (Qidwai and Chen, 2009). In application areas, we believe simplicity is attractive, based on previous experience. The well-known physical rule of conserving energy also isn't followed for image processing. As illustrated in Figure 1.1, all results obtained utilizing relatively basic processing methods in the first 10 minutes of the program may sometimes constitute 90% of work completed! The crucial insight is that themed graphics rendering ought to be implemented, while much of our knowledge is technique-based (Wong et al., 2004; Neubauer and Wallaschek, 2013)..

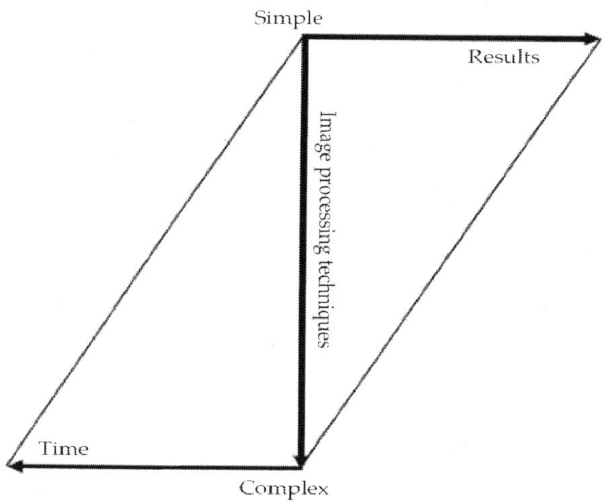

Figure 1.1. This basic graphic demonstrates how image processing works.

Source:https://www.researchgate.net/Figure/Block-diagram-of-image-processing-steps_Fig2_293376651.

A image, photograph, or any other 2D depiction of items or even a scene is usually referred to it as a image. Each image's information is presented in shades or colors. Every 2D array of integers is just a digital image. A pixel seems to be the number that indicates the luminance of each cell in a digital image, and a digital number (DN) is the number that indicates the overall brightness of each pixel (Figure 1.1) (Bourne, 2010). The digital image is made up of the data in lines and columns and therefore is stored as both a two–dimensional array. A pixel's location is determined by the row and column of its DN. Without x and y coordinates, such data is organized in such a regular pattern. Raster data refers to data that refers to coordinates. Because digital images are nothing but data arrays, arithmetic on the DNs of images is simple. Image processing refers to the mathematical procedures performed on digital images (Madruga et al., 2010).

Layers are a third level that may be added to digital image data (Figure 1.2). Images of the same subject with varying data are referred to as levels. Layers comprise images of distinct spectrum ranges termed band or channel in spectral bands imaging. Every color photograph shot with such a digital camera, for example, is made up of 3 bands, each having red, greenish, and blues spectral data. When referring to multispectral photographs, the word bands is more often used than layer (Moussavi, 2006; Ozawa et al., 2018). Geometries recorded multi-dimensional data collections of the same scene may be thought of as layers of a image in general. We could digitize a topographic map, as an example, and afterwards, co-register the digital map with a Landsat image. The map viewer, in addition to the 7 TM spectral bands, forms an extra layer of the scene. Likewise, if a digital elevation model (DEM) data is used to rectify a SPOT image, the DEM may be regarded as a level of the SPOT image in addition to its 4 spectral channels (Møller, 1992).

In just this way, a collection of updated accordingly digital images may be thought of as three-dimensional (3D) data, with both the 'thirds' dimension serving as a connection among image processing with GIS (Shi and Real, 2009). Any digital image may be saved as a document in a computer data storage on a hard disk drive, memory card, CD, or other media. It may be seen on the monitor screen in black and white (B/W) or in colors, as well as in paper version formats like films or printing. During numerical methods, it may alternatively be output as both a plain array of integers. Its benefits as just a digital image include:

- The photos could be equally reproduced without even any alteration and lack of data (Sainarayanan et al., 2007);
- These images could be mathematically treated to generate new images without altering the original images;
- Those images could be mathematically processed to obtain new photos without changing the existing images;
- Images may be sent digitally from and to faraway places without losing information.

Sensor arrays aboard airplanes or spaceships, including such meteorological satellites, collect remote sensing-based images. Sensor systems may be classified into two parts: passively sensors or active sensors. Passive multispectral advanced optical systems (Baumgarte and Faller, 2003; Canclini et al., 2014).

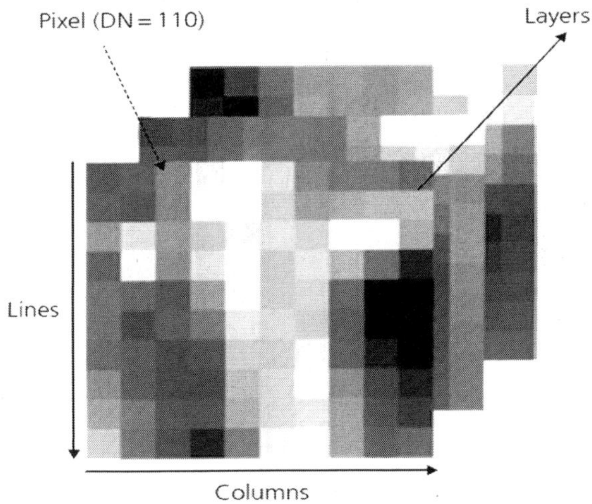

Figure 1.2. A digital image and its components.

Source: https://www.geeksforgeeks.org/digital-image-processing-basics/.

Radiation from the sun is the primary light source for imaging sensors. Cross-track and push broom multispectral scanners, and also digital cameras, are classic examples. Multispectral radar, for example, is just a sensor technology that supplies its light for imaging (SAR). Although this book does not go into considerable detail about important remote sensing data including related sensor systems, we provide a synopsis in Appendix A for your convenience (Reas and Fry, 2006).

1.2. DIGITAL IMAGE DISPLAY

We lived in a colorful world. All colors of things are the consequence of electromagnetic waves from lighting sources being selectively absorbed and reflected. The naked eye can only see light in the spectrum of 0.38–0.75 m, which is a relatively tiny portion of a sunlight spectra spectrum (Van der Voorde et al., 1986). The universe is significantly extra colorful than human eyes could see. Remote sensing data can collect across a considerably larger spectrum than human perception, and also the resulting digital images may be shown on a monitor screen either in B/W as well in color (Saxton et al., 1979).

Intensity and color represent visual representations of image data held as digital image DNS for digital image displays, although typically may not always reflect all physical meaning of these DNS. Later, when we consider fake color combinations, we'll go over this in more detail (Jensen, 1986).

The list consists of the wavelength of the primary spectral areas utilized in remotely sensed (McAndrew, 2004):

Visible light (VIS):	0.4–0.7 μm
Blue (B)	0.4–0.5 μm
Green (G)	0.5–0.6 μm
Red (R)	0.6–0.7 μm
Visible-photographic infrared:	0.5–0.9 μm
Reflective infrared (IR):	0.7–3.0 μm
Nearer infrared (NIR)	0.7–1.3 μm
Short-wave infrared (SWIR)	1.3–3.0 μm
Thermal infrared (TIR):	3–5 μm, 8–14 μm
Microwave:	0.1–100 cm

The words inside the parentheses inside the ranking above represent generally utilized spectroscopic range acronyms. The VNIR spectroscopy range, which covers visible light as well as closer infrared, is by far the most famous for internet service spectral band sensing devices (Peli, 1992).

1.2.1. Monochromatic Display

A monochrome display may show any image as just a B/W image, whether it is a panchromatic image or even a spectrum band from a multispectral image. Transforming DNS to electronic signals in a sequence of energy states generates varying gray tones (brightness) spanning black to white, resulting in a B/W visual display. Many image processing systems can show DNS ranging from 0 (black) to 255 (white) in an 8-bit graphic representation, which translates to 256 gray levels (white) (Lesem et al., 1967). The optical range of these displays is sufficient for human perception. It's also enough for several of the most regularly utilized remote sensing-based images, including such Landsat TM/ETM+, SPOT HRV, & Terra1 ASTER VIRSWIR, since their DN ranges aren't much bigger than 0–255. Several remote sensing-based photos, but on the other hand, have a far broader DN range beyond 8 bits, like Ikonos and QuickBird images, which have an 11-bit DN ranging (0–2047), and Landsat images Operational Land Imager (OLI), that has a 12-bit DN range (Bove, 2012). In this example, the images may be seen on an 8-bit monitor screen in a variety of ways, including compression the DN ranges into 8 bits or presenting the image in scenes of numerous 8-bit periods across the entire DN range. Most sensor arrays have a wide dynamic range, allowing them to collect at all levels of radiation energy without requiring localized sensor adjustments. Because the received solar energy does not fluctuate considerably within a small image scene, the scene's real dynamic range is generally much smaller than that of the sensor's complete frequency response, and so may be readily converted into an 8bit DN scope for displaying (Morgan et al., 2005; Siegel et al., 2006). The luminosity (gray level) of a pixel in some kind of monochrome screen of a different spectral image is equal to the reflected signal in these bands from either the corresponding ground region. For example, light red looks lighter than dark red in a B/W presentation of a red band image. That's also accurate for invisible wavelengths (e.g., infrared bands), even if the 'shades' really aren't visible. It seems, that each digital image is made up of DNS, whose physical interpretation varies depending on the image's source. A monochrome display renders DNS into gray tones ranging from black to white, with no regard for their physical significance (Sugimoto and Ichioka, 1985).

1.2.2. Tristimulus Color Theory and RGB (Red, Green, Blue) Color Display

If you're familiar with the construction and operation of such a color television tube, you'll understand that it's comprised of three color guns: red, greenish, and blues. These three hues are referred to as the primary color schemes. On television, based on these three fundamental colors' light may make every color. The tristimulus chroma theory can explain this characteristic of a human's color vision. The sensitivity of each type of cone in the retina image is a consequence of the frequency of received light; their maximum at 440 nm (blue), 545 nm (green), and 680 nm (red) (red). To put it another way, every cone is especially sensitive to one of the three basic colors: red, green, and blue (Yamaguchi et al., 2008).

And although visible light represents electromagnetic waves with a spectrum ranging from 380–750 nm, any person's perception of color is based on the proportion of each of these three kinds of cones activated, and may therefore be described as a tuple of integers (r, g, b). The impression of color C is formed by stimulating various regions of each cone kind with the needs of life color C light (Muthu et al., 2002):

$$C = rR + gG + bB$$

Equal amounts of such three fundamental colors (r = g = b) produce white or gray, but equal amounts of any two basic colors produce a complementary color. The complementary colors of red, green, and blue are cyan, magenta, and yellow, as seen in Figure 1.3. These three complementary colors can also be used as primary in color laser printers to create a variety of colors. If you've ever tried your hand at color paintings, you'll understand that every color could be created by combining three colors: red, yellow, and blue; the theory is the same (Ford and Roberts, 1998). The tristimulus colored theorem is used only in digital image color presentation. A colored monitor, like such a color television, is made up of three color guns that are accurately recorded in blue, green, and red. Pixels inside a image are presented in reds of varying intensities (i.e., dark red, bright red, etc.) based on the DNS in the red cannon. The green-blue firearms are in the same boat. When the red, green, and blue bands of a spectral band's images are presented in red, green, and blue at the same time, a color image is formed in which the DNS of the red, green, and blue bands determine the colors of a pixel (r, g, b). For example, a pixel with a red and green DNS of 255 and a blue DN of 0 would show on the screen as complete yellow (Ohno and Hardis, 1997). An Advanced RGB Color Composite System is a type of colored display system. Various colors are produced using additional compositions of red, green, and blue elements in this scheme. Imagine the elements of an RGB show as that of the coordinate directions of a 3D color space, as illustrated in Figures 1.4 and 1.5; the RGB color cube is defined by the maximum possible DN level for each element of the display. A vector from the source to someplace inside the color cubes may be used to describe any image pixel in this scheme. Most conventional RGB display systems can show up to 24 bits = 2563 distinct colors at 8 bits per pixel/channel (Connolly and Fleiss, 1997). This capability is sufficient to produce a 'real color image.' Although pixel vectors that fall on this line have equivalent contributions in red, green, and blue (i.e., r = g = b), the lines from the source of the color cubes to the opposing convex corner are called the gray line. Every pixel would sit just on gray lines if the red, green, and blue portions are all of the same band. Even when a transflective display system is employed, a B/W image will be created in this instance (Muthu et al., 2002).

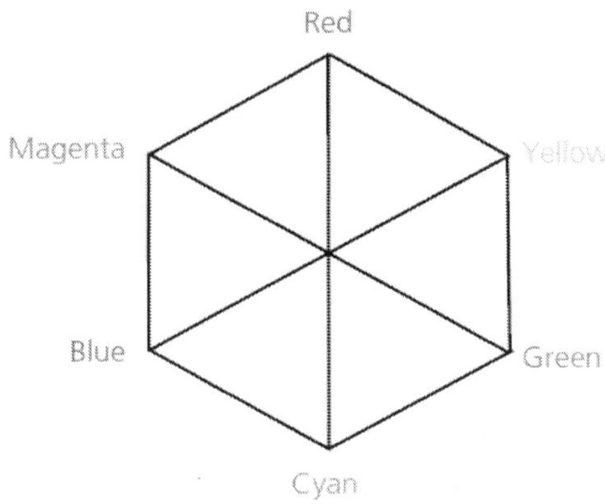

Figure 1.3. The relationship between primary and complementary colors.

Source: https://www.technologystudent.com/designpro/pricol1.htm.

Figure 1.4. Illustration of RGB additive color image presentation.

Source: https://www.dreamstime.com/stock-illustration-rgb-additive-colors-model-illustration-image53432489.

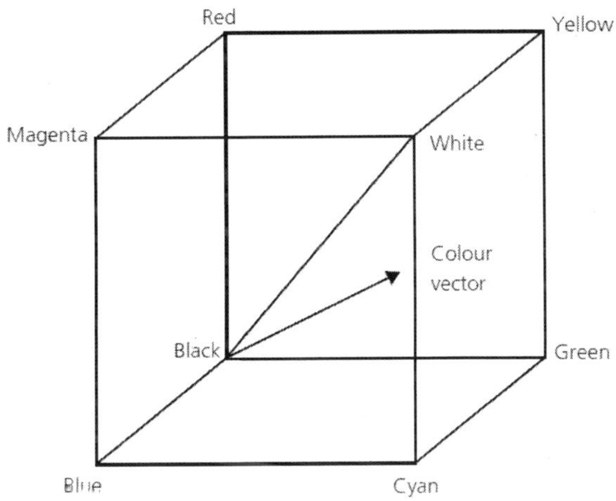

Figure 1.5. The RGB colour cube.

Source: https://www.researchgate.net/figure/the-RGB-color-cube_Fig1_304240592.

Colors are utilized as a tool for data visualization for the color presentation of all digital images, even though they're in the visible spectrum of 380–750 nm. As a result, for digital image displays, the distribution of each main color for a spectral range or layers may be based just on the user's needs, that might or might not match the actual color of the band's spectral range. A real color composite (TCC) image is created when three image bands in the red, green, and blue spectral regions are shown in RGB. A false color composites (FCC) image is created if indeed the graphic bands presented in red, green, or blues need not match the spectrum of these three fundamental colors (Wyble and Rosen, 2004).

The so-called standardized false color composite (SFCC), where the NIR band is presented in red, the red band in green, and the green band in blue, is a good example. Any flora is properly highlighted in red by the SFCC. We need a fake colored composite to present several image layers that have no spectral importance. Real color composites are a particular instance of false color composite, which is the generalized form of RGB color displays (Yoshiyama et al., 2010).

1.2.3. Pseudo-Color Display

Because the human visual system can distinguish considerably many colors from gray levels, color can be utilized to successfully emphasize minute gray-level changes in a black-and-white image. Pseudocolor presentation is a method for displaying a monochromatic image as just a color photo. Every gray grade is assigned a different color to create a pseudocolor image (Clarke and Leonard, 1989). This may be accomplished by participatory color editing or through automated transformations based on logic. Assigning a succession of gray levels to colors of excessive amount intensity and wavelength is a typical method. Pseudocolor display has both advantages and disadvantages. The serial numerical link between multiple DNS is successfully portrayed whenever a digital image is exhibited in a gray scale utilizing its DNS in a monochromic display (Twellmann et al., 2005). So because colors allocated to distinct gray levels are not statistically associated in a numeric sequence, this vital information is lost in a pseudocolor display. In a pseudocolor display, the image is indeed a image of signs; this is no longer a bitmap representation! The grayscale B/W display may be thought of as a kind of pseudocolor display where a progressive grayscale based on DN level replaces the color scheme. As demonstrated in Figure 1.6, we may often utilize a mix of B/W and pseudocolor presentation to emphasize critical information in a certain DN range in shades over a grayscale backdrop (Lavergne and Willm, 1977).

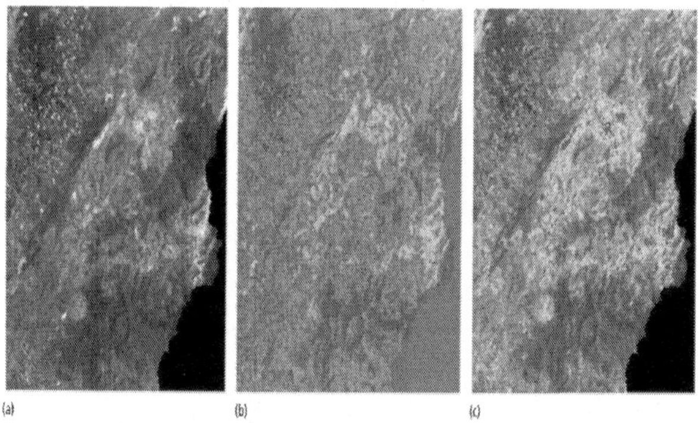

Figure 1.6. (a) A grayscale (B/W) display; (b) a pseudo color presentation of the same image; and (c) on a grayscale backdrop, the strongest DNS are highlighted in red.

Source: https://www.kdnuggets.com/2019/12/convert-rgb-image-grayscale.html.

1.3. CHARACTERISTICS OF DIGITAL IMAGE

Through some kind of sampling process known as digitizing, a digital image a[m, n] represented in a 2D discrete space is created from an analog image an (x, y) in a 2D continuous domain. Section 1.5 will go into the mathematics of that sampling process. For the time being, let's look at some fundamental terminology related to digital images. The figure depicts the impact of digitization (Fazzini et al., 2010).

N row and M column are used to partition the 2D continuous image a(x, y). A pixel is the point where a row and a column meet. a[m, n] is the value allocated to the integers values [m, n], where m=0,1,2,…,M–1 and n=0,1,2,…,N–1 (Liu et al., 2009).

In most situations, an (x, y), which we can think of as the measurable signal that intrudes just in front of a 2D sensor, is a function of the number of factors, including depth (z), color (), or time (t) (Kim & Kim, 2013).

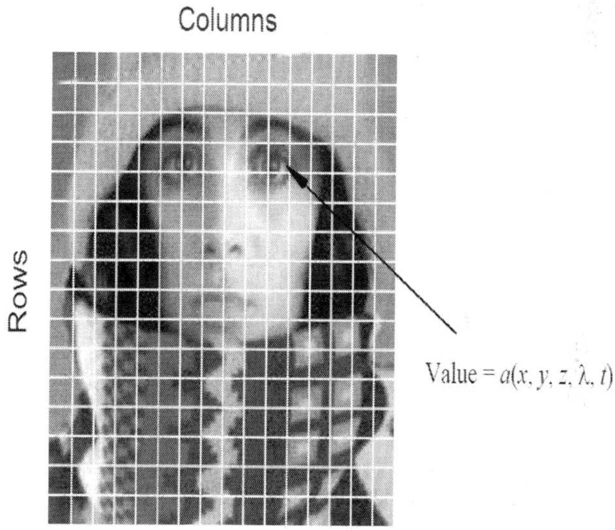

Figure 1.7. Digitization of a continuous image. The integer brightness value 110 is assigned to the pixel at coordinates [m=10, n=3].

Source: emanticscholar.org/paper/Digital-image-restoration-by-Wiener-filter-in-2D-Khireddine-Benmahammed/c112e05124b-d594a1a7209e4b23b2b99dd06e35a/figure/0.

Figure 1.7 shows the image split into N = 16 row and M = 16 column. The mean intensity in each pixel is adjusted to the closest numerical value and given to each pixel. Intensity quantized, or simply quantization is the technique of encoding the intensity of a 2D signal at the given coordinate is an integer number with L distinct gray levels (Tucker and Chakraborty, 1997).

1.3.1. Common Values

The different parameters found in image processing have standards. Video standards, algorithmic needs, or even the determination to maintain digital circuits simple may all influence these numbers (Besag, 1989) (Table 1.1).

Table 1.1. Common Values of Digital Image Parameters

Parameter	*Symbol*	*Typical values*
Rows	N	256,512,525,625,1024,1035
Columns	M	256,512,768,1024,1320
Gray Levels	L	2,64,256,1024,4096,16384

M=N=2K where K = 8,9,10 is a very common situation. Digital circuits or the usage of algorithms like the (rapid) Fourier transform may be used to justify this (Choi and Kim, 2005).

L=2B, wherein B is the unit of data in the binary encoding of the brightness, is the typical number of different gray levels. When B>1, we're talking about a gray-level image, and when B=1, we're talking about a source image. There are just two gray levels in a binary image, that are alluded to as "black" and "white" or "0" and "1" respectively (Abu Dalhoum et al., 2012).

1.3.2. Characteristics of Image Operations

Imaging operations may be classified and described in several ways. The goal is to figure out what kind of outcomes we can anticipate from a certain type of operation and how much computing work it will take (Bossuyt, 2013).

1.3.2.1. Types of Operations

As indicated in Table 1.2, three kinds of operations may be done to digital images to change an input image a[m, n] into an output image b[m, n] (or another representation) (Sencar and Memon, 2009).

Table 1.2. Types of Image Operations. Image size = N ′ N; Neighborhood size = P × P. Note that the Complexity is Specified in Operations Per Pixel

Operation	Characterization	Generic Complexity/Pixel
• *Point*	– the output value at a specific coordinate is dependent only on the input value at that same coordinate.	*constant*
• *Local*	– the output value at a specific coordinate is dependent on the input values in the *neighborhood* of that same coordinate.	P^2
• *Global*	– the output value at a specific coordinate is dependent on all the values in the input image.	N^2

This is shown graphically in Figure 1.8.

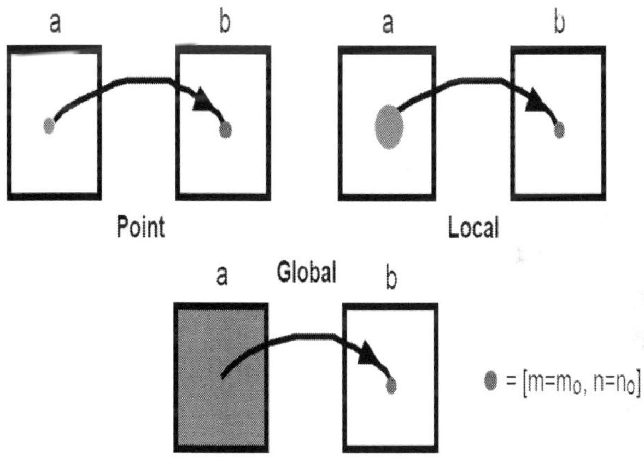

Figure 1.8. Illustration of several sorts of image manipulations.

Source:https://www.officinaturini.com/files/ImageProcessingFundamentals/noframes/fip-Characte.html.

1.3.2.2. Types of Neighborhoods

In today's image processing, neighborhood operations are crucial. It's important to understand how images may be collected and how it connects to the many neighborhoods that can be utilized to analyze a image (Oberholzer et al., 1996).

- Rectangle sampling—Images are surveyed in most circumstances by putting a rectangle grid across them, as seen in Figure 1.10. As a consequence, the sort of sample represented in Figure ab is produced.
- Hexagon shaped sampling—Figure 3c depicts an alternate sample strategy known as hexagon sampling. These sampling strategies have indeed been thoroughly investigated, and both offer a potential regular tiling of continuous image space. Therefore, we will limit our discussion to rectangle sampling since it is still the preferred approach owing to equipment/software constraints (Henry, 2013).

Depending on the image pixels in the vicinity of a[m=mo, n=no], localized operations yield an output pixel value b[m=mo, n=no]. The 8-connected and 4-connected neighborhoods in a rectangular sample and the 6-connected neighborhood in a hexagonal sample are two of the most prevalent neighborhoods, as shown in figure (Peters, 2014).

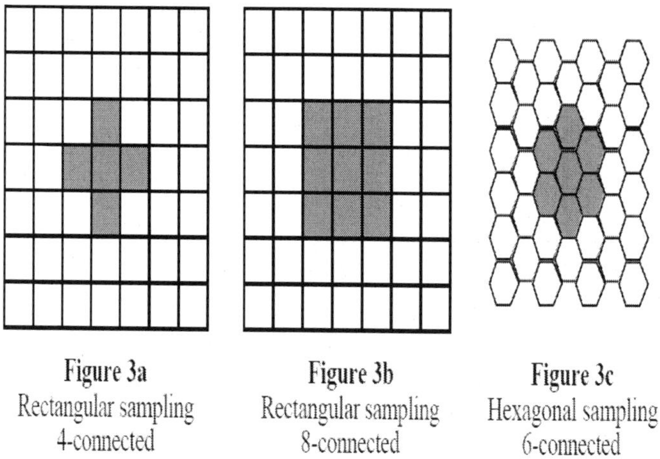

Figure 1.9. Illustration of common neighborhood.

Source:https://www.gettyimages.com/detail/illustration/neighborhood-similar-homes-drawing-royalty-free-illustration/506065602.

1.4. SOUND

Audio makes up a large portion of the information we collect and interpret daily. If we have a face-to-face discussion with somebody and listen to the noises inside a forest or even a street, the majority of these sounds are transported straight from the source to our ears (Munoz et al., 2016). Speaker systems in many types of audio devices, such as mobile phones, digital music players, home theaters, radios, and tv sets, pro These devices' noises are either created by stored information within the device, or electromagnetic frequencies that managed to be picked up by such an antenna analyzed, and then transformed to sound. This is the type of sound we'll be looking at in this section. Electronic sound is indeed the representation of sound that is stored within the equipment or managed to be picked up by antennas (Busciglio et al., 2008). This has several restrictions, but it also makes manipulating and processing sounds on a computer relatively simple. duce a significant portion of the noises. The physical processes of tiny fluctuations in air density around the ears correlate to what we experience as sound. Larger fluctuations correlate to louder noises, whereas quicker changes correspond to higher-pitched sounds. Atmospheric air pressure fluctuates continually throughout time, yet it has an exact value at any particular instant. As a result, sound may be thought of as a mathematical equation (Wang et al., 2013).

We would quickly explore the essential qualities of sounds in the subsequent subsections: first, the relevance of the magnitude of the fluctuations, and afterwards the number of changes per second, or the frequencies of audio (Jin et al., 2015). We also take into account the notion that every good music may be constructed from extremely basic base sounds. Because a sound may be thought of as a functional, any reasonable function can be built from extremely basic foundation functions (Gloe and Böhme, 2010). Fourier analysis is the theoretical study of this, and we'll look at it from a practical and computing standpoint in the following chapters (Li et al., 2022).

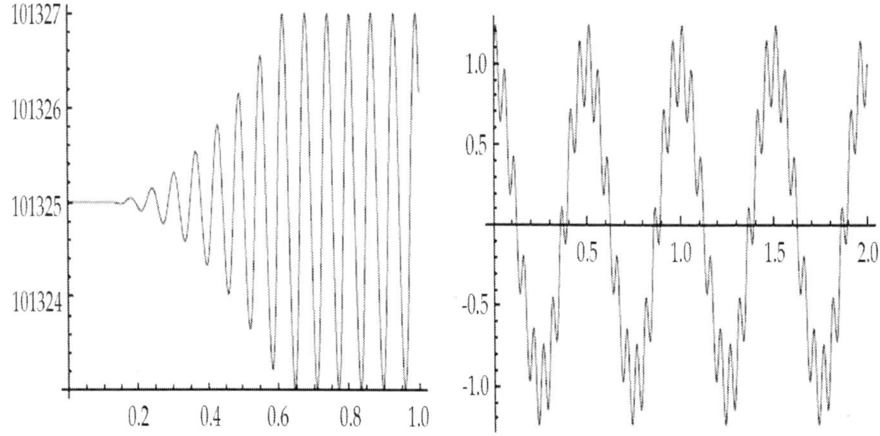

(a) A sound shown in terms of air pressure (b) A sound shown in terms of the difference from the ambient air pressure

Figure 1.10. Two examples of audio signals.

Source: https://www.researchgate.net/figure/Two-independently-generated-examples-of-audio-source-signals-for-testing-the-validity-of_Fig1_3451138.

In coming chapters we also consider the basics of digital audio, and illustrate its power by performing some simple operations on digital sounds. (Rajput et al., 2018).

1.4.1. Sound Pressure and Decibels

In figure 1.11, the waves in air density are exhibited over time as an example of a basic sound. We could see that the original air pressure is 101 325 (we'll get back to a unit later), but then the pressure continues to fluctuate even more until it oscillates consistently between 101 323 and 101 327. There'll be no sound in the region in which the pressure is consistent, however as the fluctuations grow in magnitude, the sounds will get louder and louder until around time t = 0.6 when the frequency of the oscillations reaches constant. Below is a quick rundown of some fundamental air pressure information (Krothapalli et al., 1999).

At a standard air pressure of 101 325 Pa, the audio was captured. After the sound began, the pressure began to fluctuate either behind or in front of this number, and after a brief transitory phase, the pressure constantly fluctuated between 101 324 Pa and 101 326 Pa, corresponding to 1 Pa changes around the fixed amount. Common noises correlate to pressurization fluctuations of around 0.00002–2 Pa, but a fighter jet may create variations of up to 200 Pa (Zhu et al., 2004). Ear damage may occur after a brief exposure to pressure changes of around 20 Pa. In 1883, an eruption in Krakatoa, Indonesia, generated a sound wave with fluctuations of about 100 000 Pa, that could be heard from 5000 kilometers distant (Keele Jr, 1974).

Because most people are solely concerned with variations in air pressure whenever addressing sound, the background air pressure is eliminated from the measure. This is equivalent to removing 101 325 from the vertical axis numbers in Figure 1.10(a). The deduction for another sound is shown in Figure 1.10(b), and we can see that the audio has a gradual, cos-like fluctuation in air density, with some smaller and quicker variations imposed on it (Miller, 2006). The combination of numerous types of regular air pressure oscillations is characteristic of broad sounds. The magnitude of the oscillations is proportional to a sound's volume. We've found that changes in audible noises may vary anywhere from 0.00002 Pa to 100 000 Pa. Because of this vast range, it is typical to use a logarithmic scale to quantify the loudness of a sound. Air pressure is often standardized to a value between and 1: The ambient air pressure is represented by 0; the smallest and maximum computable air pressures are represented by 1 & 1 (Fox and Ramig, 1997),

The following data boxes simplify and explain the logarithm decibels scale, as well as the prior description of what a sound is. The cube of the noise level is included in the concept of Lp because it indicates the sound's strength, which is important for our perception of loudness. Figure 1.10 shows synthetic noises, which were created using mathematical methods. The noises in Figure 1.11, on either hand, depict variations in air pressure when no mathematical equation is used, like when a song is played. There are so many oscillations in (a) that the intricacies are hard to detect, however, if we close in as in (c), we can see there is a constant underlying all ink. It's vital to remember that, even over a short amount of time, air pressure changes far more than this (c). Unfortunately, the measurement was unable to detect such changes, and we are unsure if we'd be able to detect such quick changes (Guinan Jr, & Peake, 1967).

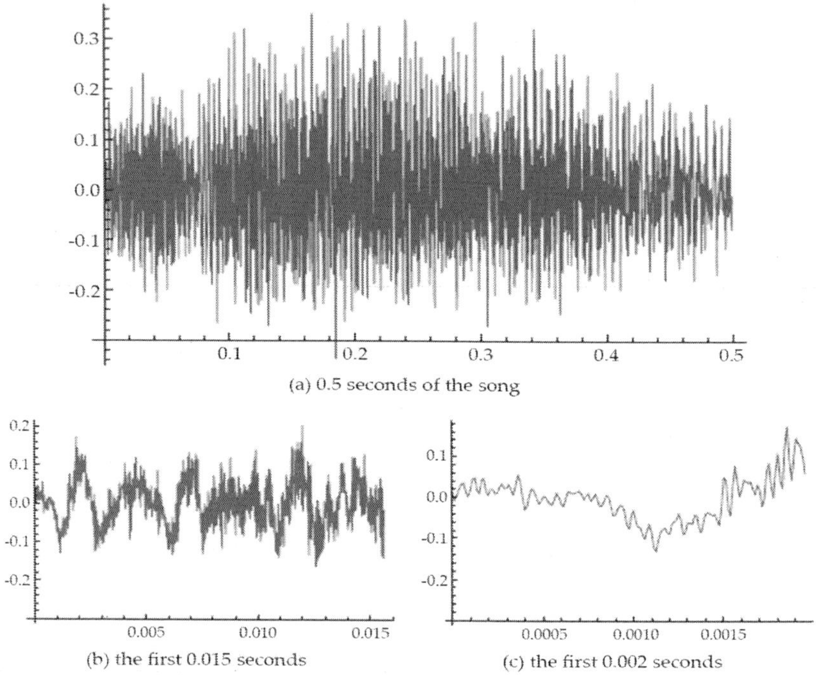

Figure 1.11. Variations in air pressure during parts of a song.

Source: https://www.quora.com/What-is-the-variation-of-pressure-in-the-Earths-atmosphere.

1.4.2. The Pitch of a Sound

A voice may have another essential attribute besides the amplitude of the fluctuations in air pressure: the rate (speed) of the changes. The speed of fluctuations in air pressure fluctuates throughout most sounds, but they should fall within a certain range if we are to interpret them as sound. To clarify these ideas, first, consider just what means for such a function to be periodic (Reinhard et al., 2002).

Notice that even if f(t) is known for every t in the range [0,], then all variables of a periodical function f of periods are known. The trigonometric functions are the prototypes of periodic functions, and we're especially interested in sin t and cos t. Because sin(t + 2) = sint, we can see that sin t has a period of 2 and cos*t* has a period of 2 (Reinhard et al., 2006).

A basic method for changing the duration of a periodic wave is to multiply the parameter by the constant.

When t fluctuates in the range [0, 2], the function in (a) is the simple sin t, which spans one interval. Every period is compressed into the range [0, 1] by doubling the input by 2, giving the functional sin(2t) a frequency of v = 1. Then, by doubling the parameter by two, we squeeze two full periods into the range [0, 1], giving the function sin(22t) a frequency of 2 (Cederberg et al., 1999). The arguments have already been divided by 5 in (d), resulting in a frequency of 5 and a total of five periods inside the interval [0, 1]. Regardless of the value of any function of the type sin(2t + a) has a frequency. Because sound may be represented by functions, a sound with speed can be described as a right triangle with speed (Ahmad et al., 2012) (Figure 1.12).

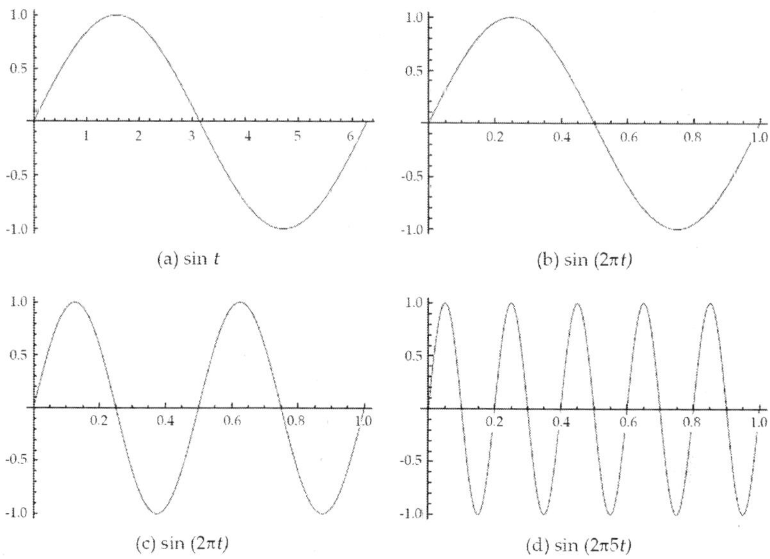

Figure 1.12. Versions of sin with different frequencies.

Source: https://musiccms.ucsd.edu/404/index.html?url=http%3A%2F%2Fmusicweb.ucsd.edu%2F~trsmyth%2FaddSynth171%2FAdding_Sinusoids_at.html.

This is how a clear tone with such a frequency of 440 Hz sounds, and that's how a clean tone with such a frequency of 1500 Hz sounds. Every voice may be categorized as a function. Any sensible function may be written as a sum of simple sin- and cos- curves with integer frequencies, as we'll see in the following chapter. When we convert this into sound qualities, we get an essential concept (Yamauchi et al., 2012).

The most fundamental implication is that it teaches us how to construct every sound from the simplest basic building block of pure–tone. This implies that, rather than storing f itself, we may save the frequency components of a sound f (Dutta et al., 2013). This also opens up the prospect of lossless digital sound compaction: It comes out that in normal audio transmission, the frequency bands contain the most information, whereas certain frequencies are virtually empty. If we modify the frequency with modest contributions a very little bit and convert them to 0, and then save the signal by just recording the nonzero component of the frequency components, we can use this for compressing. When the audio is played back, we first use an inverted mapping to transfer the modified numbers to the modified frequency components back to a regular functional representation (Yokota and Teale, 2014). This lossy image compression method is basically what commercialized audio formats utilize in reality. Commercial software, on the other hand, accomplishes it all in a more complex manner, resulting in higher compression rates (Cahill and McGill-Franzen, 2013).

It is simple to make a voice from a mathematical equation with suitable software; we may 'play' the expression. We heard a pleasant sound with a fairly distinct pitch when we perform functions like sin(2440t), as predicted. There are, however, several additional methods for a function to oscillate predictably. For example, the functions in Figure 1.10(b) oscillate Two times per second, however, does not have a speed of 2 Hz because it is not a clear tone. This sound isn't nice to listen to either. We'll look at two more key instances of this, both are not flat trigonometry functions (Cohen, 1993).

1.5. DIGITAL SOUND

We looked at some fundamental features for sounds in the last part, however, all was in terms of its functionality created for all time occurrences in some interval. Sound is frequently digital on computers and other types of media players, which indicates that the sounds are expressed by a large number of function values rather than a function specified for all periods in a certain interval (Gloe and Böhme, 2010).

Digital audio is often taken from an analog (continuous) audio stream. Pulse Code Modulation is a common method for doing this (PCM). During frequent intervals, the sound stream is sampled, as well as the selected sample values are saved in an appropriate numerical format. For various types of audio, the sampling rate, as well as the precision of numbers formats used to store the sample, could change, both of which have an impact just on audio quality (Busciglio et al., 2008).

The amount of bit / s, which is the sum of a sample rate and the number of bits (binary digits) used to record each sample, is frequently used to gauge quality. The data rate is another term for this. Sampling must be kept inside a folder or in ram on a computer for that to be capable of playing a digitized sound. Digital audio formats are utilized to achieve this efficiently (Lee and Shinozuka, 2006).

Every program may do simple operations and calculations with multichannel audio. Let's have a glance at how we might do this. The frame rate fs is combined with an array of sample data x = (xi)N 1 i=0 to create digital audio. Performing activities on sound, therefore, entails conducting the necessary calculations using the data samples and sampling frequency. Producing audio is the most fundamental action we can do on it, and if we're dealing with sound, we'll need a method to do so (Rajput et al., 2018).

In the previous part, you may well have heard clear tones, square waves, and triangular waves. The accompanying audio files are generated in a manner that we shall discuss momentarily, then stored in an online directory and connected to such notes. Once you click on such files, the software on your computer was capable of playing them. We'll now show you how to produce the very same sounds in MATLAB. The two functionalities are now in place (Cheddad et al., 2010).

```
playblocking(playerobj)
playblocking(playerobj,[start stop])
```

These just play a portion of audio wrapped by playerob object (we will shortly see how we can construct such an object from given audio samples and sampling rate). The technique that is generating the audio will be blocked until someone has stopped playing. We'll come back to this feature later because we can play sounds in reverse order. The very first action plays the complete audio portion (Ringrose et al., 2013).

The replay of the second procedure begins at sampling start and stops at sample stops. These functionalities are simply software connections to the computer's device. It simply provides an array of sound samples at a sample rate to the sound system, which reconstructs the sound into an analog sound signal using some algorithm. The sound is subsequently transmitted to the amplifiers through this analog signal (Wang et al., 2012).

REFERENCES

1. Abu, D. A. L., Mahafzah, B. A., Awwad, A. A., Aldhamari, I., Ortega, A., & Alfonseca, M., (2012). Digital image scrambling using 2D cellular automata. *IEEE Multimedia* (4th edn., pp. 2–6).
2. Ahmad, K., Yan, Y., & Bless, D., (2012). Vocal fold vibratory characteristics of healthy geriatric females—Analysis of high-speed digital images. *Journal of Voice, 26*(6), 751–759.
3. Baumgarte, F., & Faller, C., (2003). Binaural cue coding-part I: Psychoacoustic fundamentals and design principles. *IEEE Transactions on Speech and Audio Processing, 11*(6), 509–519.
4. Besag, J., (1989). Digital image processing: Towards Bayesian image analysis. *Journal of Applied Statistics, 16*(3), 395–407.
5. Bossuyt, S., (2013). Optimized patterns for digital image correlation. In: *Imaging Methods for Novel Materials and Challenging Applications* (Vol. 3, pp. 239–248). Springer, New York, NY.
6. Bourne, R., (2010). *Fundamentals of Digital Imaging in Medicine* (2nd edn., pp. 1–5). Springer Science & Business Media.
7. Bove, V. M., (2012). Display holography's digital second act. *Proceedings of the IEEE, 100*(4), 918–928.
8. Busciglio, A., Vella, G., Micale, G., & Rizzuti, L., (2008). Analysis of the bubbling behavior of 2D gas solid fluidized beds: Part I. Digital image analysis technique. *Chemical Engineering Journal, 140*(1–3), 398–413.
9. Cahill, M., & McGill-Franzen, A., (2013). Selecting "app" ealing and "app" ropriate book apps for beginning readers. *The Reading Teacher, 67*(1), 30–39.
10. Canclini, A., Markovic, D., Bianchi, L., Antonacci, F., Sarti, A., & Tubaro, S., (2014). A robust geometric approach to room compensation for sound field rendering. *IEICE Transactions on Fundamentals of Electronics, Communications and Computer Sciences, 97*(9), 1884–1892.
11. Cederberg, R. A., Frederiksen, N. L., Benson, B. W., & Shulman, J. D., (1999). Influence of the digital image display monitor on observer performance. *Dentomaxillofacial Radiology, 28*(4), 203–207.
12. Cheddad, A., Condell, J., Curran, K., & Mc Kevitt, P., (2010). Digital image steganography: Survey and analysis of current methods. *Signal Processing, 90*(3), 727–752.

13. Choi, K. Y., & Kim, S. S., (2005). Morphological analysis and classification of types of surface corrosion damage by digital image processing. *Corrosion Science, 47*(1), 1–15.
14. Clarke, F. J. J., & Leonard, J. K., (1989). Proposal for a standardized continuous pseudo-color spectrum with optimal visual contrast and resolution. In: *Third International Conference on Image Processing and its Applications* (Vol. 1, pp. 687–691). IET.
15. Cohen, M., (1993). Throwing, pitching and catching sound: Audio windowing models and modes. *International Journal of Man-Machine Studies, 39*(2), 269–304.
16. Connolly, C., & Fleiss, T., (1997). A study of efficiency and accuracy in the transformation from RGB to CIELAB color space. *IEEE Transactions on Image Processing, 6*(7), 1046–1048.
17. Deserno, T. M., (2010). Fundamentals of biomedical image processing. In; *Biomedical Image Processing* (Vol. 1, pp. 1–51). Springer, Berlin, Heidelberg.
18. Dutta, S., Pal, S. K., Mukhopadhyay, S., & Sen, R., (2013). Application of digital image processing in tool condition monitoring: A review. *CIRP Journal of Manufacturing Science and Technology, 6*(3), 212–232.
19. Fazzini, M., Mistou, S., Dalverny, O., & Robert, L., (2010). Study of image characteristics on digital image correlation error assessment. *Optics and Lasers in Engineering, 48*(3), 335–339.
20. Ford, A., & Roberts, A., (1998). *Color Space Conversions* (Vol. 1, pp. 1–31). Westminster University, London.
21. Fox, C. M., & Ramig, L. O., (1997). Vocal sound pressure level and self-perception of speech and voice in men and women with idiopathic Parkinson disease. *American Journal of Speech-Language Pathology, 6*(2), 85–94.
22. Gloe, T., & Böhme, R., (2010). The 'Dresden image database' for benchmarking digital image forensics. In: *Proceedings of the 2010 ACM Symposium on Applied Computing* (Vol. 1, pp. 1584–1590).
23. Guinan, Jr. J. J., & Peake, W. T., (1967). Middle-ear characteristics of anesthetized cats. *The Journal of the Acoustical Society of America, 41*(5), 1237–1261.
24. Henry, C. J., (2013). Metric free nearness measure using description-based neighborhoods. *Mathematics in Computer Science, 7*(1), 51–69.

25. Jensen, J. R., (1986). *Introductory Digital Image Processing: A Remote Sensing Perspective* (2nd edn., pp. 3–6). Univ. of South Carolina, Columbus.
26. Jin, W., Liu, J., Wang, Z., Wang, Y., Cao, Z., Liu, Y., & Zhu, X., (2015). Sound absorption characteristics of aluminum foams treated by plasma electrolytic oxidation. *Materials, 8*(11), 7511–7518.
27. Keele, Jr. D. B., (1974). Low-frequency loudspeaker assessment by nearfield sound-pressure measurement. *Journal of the Audio Engineering Society, 22*(3), 154–162.
28. Kim, S. W., & Kim, N. S., (2013). Dynamic characteristics of suspension bridge hanger cables using digital image processing. *NDT & E International, 59*(1), 25–33.
29. Korpel, A., (1981). Acousto-optics—A review of fundamentals. *Proceedings of the IEEE, 69*(1), 48–53.
30. Krothapalli, A., Rajkuperan, E., Alvi, F., & Lourenco, L., (1999). Flow field and noise characteristics of a supersonic impinging jet. *Journal of Fluid Mechanics, 392*(1), 155–181.
31. Kunduk, M., Yan, Y., Mcwhorter, A. J., & Bless, D., (2006). Investigation of voice initiation and voice offset characteristics with high-speed digital imaging. *Logopedics Phoniatrics Vocology, 31*(3), 139–144.
32. Kuo, S. M., Lee, B. H., & Tian, W., (2013). *Real-Time Digital Signal Processing: Fundamentals, Implementations and Applications* (Vol. 2, No. 1, pp. 2–9). John Wiley & Sons.
33. Lavergne, M., & Willm, C., (1977). Inversion of seismograms and pseudo velocity logs. *Geophysical Prospecting, 25*(2), 231–250.
34. Lee, J. J., & Shinozuka, M., (2006). Real-time displacement measurement of a flexible bridge using digital image processing techniques. *Experimental Mechanics, 46*(1), 105–114.
35. Lesem, L. B., Hirsch, P. M., & Jordan, Jr. J. A., (1967). Holographic display of digital images. In: *Proceedings of the Fall Joint Computer Conference* (Vol. 1, pp. 41–47).
36. Li, X., Gao, J., Du, H., Jia, J., Zhao, X., & Ling, T., (2022). Relationship between the void and sound absorption characteristics of epoxy porous asphalt mixture based on CT. *Coatings, 12*(3), 328.
37. Liu, X., Song, W., & Zhang, J., (2009). Extraction and quantitative analysis of microscopic evacuation characteristics based on digital

image processing. *Physica A: Statistical Mechanics and its Applications, 388*(13), 2717–2726.

38. Madruga, F. J., Ibarra-Castanedo, C., Conde, O. M., López-Higuera, J. M., & Maldague, X., (2010). Infrared thermography processing based on higher-order statistics. *NDT & E International, 43*(8), 661–666.
39. Mannan, M. A., Kassim, A. A., & Jing, M., (2000). Application of image and sound analysis techniques to monitor the condition of cutting tools. *Pattern Recognition Letters, 21*(11), 969–979.
40. McAndrew, A., (2004). *An Introduction to Digital Image Processing with MATLAB Notes for SCM2511 Image Processing* (Vol. 1, pp. 1–9). 1 Semester 1.
41. Miller, P. J., (2006). Diversity in sound pressure levels and estimated active space of resident killer whale vocalizations. *Journal of Comparative Physiology A, 192*(5), 449–459.
42. Møller, H., (1992). Fundamentals of binaural technology. *Applied Acoustics, 36*(3, 4), 171–218.
43. Morgan, J. E., Sheen, N. J. L., North, R. V., Choong, Y., & Ansari, E., (2005). Digital imaging of the optic nerve head: Monoscopic and stereoscopic analysis. *British Journal of Ophthalmology, 89*(7), 879–884.
44. Moussavi, Z., (2006). Fundamentals of respiratory sounds and analysis. *Synthesis Lectures on Biomedical Engineering, 1*(1), 1–68.
45. Müller, M., (2015). *Fundamentals of Music Processing: Audio, Analysis, Algorithms, Applications* (Vol. 5, pp. 2–9). Cham: Springer.
46. Munoz, H., Taheri, A., & Chanda, E. K., (2016). Pre-peak and post-peak rock strain characteristics during uniaxial compression by 3D digital image correlation. *Rock Mechanics and Rock Engineering, 49*(7), 2541–2554.
47. Muthu, S., Schuurmans, F. J., & Pashley, M. D., (2002). Red, green, and blue LEDs for white light illumination. *IEEE Journal of Selected Topics in Quantum Electronics, 8*(2), 333–338.
48. Muthu, S., Schuurmans, F. J., & Pashley, M. D., (2002). Red, green, and blue LED based white light generation: Issues and control. In: *Conference Record of the 2002 IEEE Industry Applications Conference: 37th IAS Annual Meeting (Cat. No. 02CH37344)* (Vol. 1, pp. 327–333). IEEE.

49. Neubauer, M., & Wallaschek, J., (2013). Vibration damping with shunted piezoceramics: Fundamentals and technical applications. *Mechanical Systems and Signal Processing, 36*(1), 36–52.
50. Oberholzer, M., Östreicher, M., Christen, H., & Brühlmann, M., (1996). Methods in quantitative image analysis. *Histochemistry and Cell Biology, 105*(5), 333–355.
51. Ohno, Y., & Hardis, J. E., (1997). Four-color matrix method for correction of tristimulus colorimeters. In: *Color and Imaging Conference* (Vol. 1997, No. 1, pp. 301–305). Society for Imaging Science and Technology.
52. Ozawa, K., Koshimizu, Y., Morise, M., & Sakamoto, S., (2018). Separation of two sound sources in the same direction by image signal processing. In: *2018 IEEE 7th Global Conference on Consumer Electronics (GCCE)* (Vol. 1, pp. 511, 512). IEEE.
53. Peli, E., (1992). Display nonlinearity in digital image processing for visual communications. *Optical Engineering, 31*(11), 2374–2382.
54. Peters, J. F., (2014). Topology of digital images: Basic ingredients. In: *Topology of Digital Images* (Vol. 1, pp. 1–75). Springer, Berlin, Heidelberg.
55. Qidwai, U., & Chen, C. H., (2009). *Digital Image Processing: An Algorithmic Approach with MATLAB* (Vol. 4, No. 2, pp. 5–8). Chapman and Hall/CRC.
56. Rajput, S. K., Matoba, O., & Awatsuji, Y., (2018). Characteristics of vibration frequency measurement based on sound field imaging by digital holography. *OSA Continuum, 1*(1), 200–212.
57. Reas, C., & Fry, B., (2006). Processing: Programming for the media arts. *AI & Society, 20*(4), 526–538.
58. Reinhard, E., Stark, M., Shirley, P., & Ferwerda, J., (2002). Photographic tone reproduction for digital images. In: *Proceedings of the 29th Annual Conference on Computer Graphics and Interactive Techniques* (Vol. 1, pp. 267–276).
59. Ringrose, J., Harvey, L., Gill, R., & Livingstone, S., (2013). Teen girls, sexual double standards and 'sexting': Gendered value in digital image exchange. *Feminist Theory, 14*(3), 305–323.
60. Sainarayanan, G., Nagarajan, R., & Yaacob, S., (2007). Fuzzy image processing scheme for autonomous navigation of human blind. *Applied Soft Computing, 7*(1), 257–264.

61. Saxton, W. O., Pitt, T., & Horner, M., (1979). Digital image processing: The semper system. *Ultramicroscopy, 4*(3), 343–353.s.

62. Schreiber, W. F., (2012). *Fundamentals of Electronic Imaging Systems: Some Aspects of Image Processing* (Vol. 15, pp. 4–8). Springer Science & Business Media.

63. Schueler, C. F., Lee, H., & Wade, G., (1984). Fundamentals of digital ultrasonic processing. *IEEE Transactions on Sonics and Ultrasonics, 31*(4), 195–217.

64. Sencar, H. T., & Memon, N., (2009). Overview of state-of-the-art in digital image forensics. *Algorithms, Architectures and Information Systems Security* (2nd edn., pp. 325–347).

65. Shi, Y., & Real, F. D., (2009). Smart cameras: Fundamentals and classification. In: *Smart Cameras* (Vol. 1, pp. 19–34). Springer, Boston, MA.

66. Siegel, E., Krupinski, E., Samei, E., Flynn, M., Andriole, K., Erickson, B., & Pisano, E. D., (2006). Digital mammography image quality: Image display. *Journal of the American College of Radiology, 3*(8), 615–627.

67. Smith, J., & Finnoff, J. T., (2009). Diagnostic and interventional musculoskeletal ultrasound: Part 1. Fundamentals. *PM&R, 1*(1), 64–75.

68. Sugimoto, S. A., & Ichioka, Y., (1985). Digital composition of images with increased depth of focus considering depth information. *Applied Optics, 24*(14), 2076–2080.

69. Tohyama, M., Koike, T., & Bartram, J. F., (2000). *Fundamentals of Acoustic Signal Processing 470-471.*

70. Toriwaki, J., & Yoshida, H., (2009). *Fundamentals of Three-Dimensional Digital Image Processing* (Vol. 1. pp. 5–9). Springer Science & Business Media.

71. Tucker, C. C., & Chakraborty, S., (1997). Quantitative assessment of lesion characteristics and disease severity using digital image processing. *Journal of Phytopathology, 145*(7), 273–278.

72. Twellmann, T., Lichte, O., Saalbach, A., Wismüller, A., & Nattkemper, T. W., (2005). An adaptive extended color scale for comparison of pseudo coloring techniques for DCE-MRI Data. In: *Bildverarbeitung für die Medizin* (Vol. 1, pp. 312–316). Springer, Berlin, Heidelberg.

73. Van, D. V. F., Arenson, R., Kundel, H., Miller, W., Epstein, D., Gefter, W., & Khalsa, S., (1986). Development of a physician-friendly digital image display console. In: *Application of Optical Instrumentation in Medicine XIV and Picture Archiving and Communication Systems* (Vol. 626, pp. 541–548). International Society for Optics and Photonics.

74. Wang, W., Mottershead, J. E., Siebert, T., & Pipino, A., (2012). Frequency response functions of shape features from full-field vibration measurements using digital image correlation. *Mechanical Systems and Signal Processing, 28*(1), 333–347.

75. Wang, Y., Zhang, C., Ren, L., Ichchou, M., Galland, M. A., & Bareille, O., (2013). Influences of rice hull in polyurethane foam on its sound absorption characteristics. *Polymer Composites, 34*(11), 1847–1855.

76. Wong, Y. L., Burg, J., & Strokanova, V., (2004). Digital media in computer science curricula. In: *Proceedings of the 35th SIGCSE Technical Symposium on Computer Science Education* (Vol. 1, pp. 427–431).

77. Wyble, D. R., & Rosen, M. R., (2004). Color management of DLP™ projectors. In: *Color and Imaging Conference* (Vol. 12, pp. 228–232). Society of Imaging Science and Technology.

78. Yamaguchi, M., Haneishi, H., & Ohyama, N., (2008). Beyond red–green–blue (RGB): Spectrum-based color imaging technology. *Journal of Imaging Science and Technology, 52*(1), 10201–10205.

79. Yamauchi, A., Imagawa, H., Yokonishi, H., Nito, T., Yamasoba, T., Goto, T., & Tayama, N., (2012). Evaluation of vocal fold vibration with an assessment form for high-speed digital imaging: Comparative study between healthy young and elderly subjects. *Journal of Voice, 26*(6), 742–750.

80. Yokota, J., & Teale, W. H., (2014). Picture books and the digital world: Educators making informed choices. *The Reading Teacher, 67*(8), 577–585.

81. Yoshiyama, K., Teragawa, M., Yoshida, A., Tomizawa, K., Nakamura, K., Yoshida, Y., & Ohta, N., (2010). 29.2: Power-saving: A new advantage of multi-primary color displays derived by numerical analysis. In: *SID Symposium Digest of Technical Papers* (Vol. 41, No. 1, pp. 416–419). Oxford, UK: Blackwell Publishing Ltd.

82. Zhu, C., Liu, G., Yu, Q., Pfeffer, R., Dave, R. N., & Nam, C. H., (2004). Sound assisted fluidization of nanoparticle agglomerates. *Powder Technology, 141*(1, 2), 119–123.

CHAPTER 2

IMAGE PROCESSING BY COMBINING HUMAN AND MACHINE COMPUTING

CONTENTS

2.1. INTRODUCTION

During artificial or natural disasters, the social media platforms usages as Facebook and Twitter have considerably expanded (Hughes and Palen, 2009). People share a wide range of information, including photographs, videos, and text messages. The importance and usage of the internet data for humanitarian groups dealing with disaster management and the response have been documented in several studies. Furthermore, for crisis management and response activities, the bulk of such studies have relied almost entirely on textual material (e.g., tweets, messages, postings, etc.) (Elbassuoni et al., 2013). In contrast to the current research on the use of social media text information for crisis response, this research concentrates on the usage of social media visual material (like photographs) to demonstrate its value for disaster management and response to humanitarian groups. Data obtained from social media images, if handled quickly and efficiently, may help with fast decisions as well as other humanistic rescue operations, like obtaining situational awareness or categorization and evaluating the extent of damage throughout a disaster (Ofli et al., 2016).

In contrast to the convenience with which images may be obtained from multiple social media platforms, evaluating the vast number of picture content released after a catastrophic disaster remains a difficult task. Human annotators are often sought by humanitarian groups to mark areas of importance (e.g., destroyed shelters and obstructed roadways) in photographs. For social media filtration, the United Nations Office for the Coordination of Humanitarian Affairs (UN OCHA) uses individuals from the Digital Humanitarian Network (Ludwig et al., 2015). The usage of a machine learning technique and hybrid crowdsourcing to quickly analyze massive amounts of picture information for disaster response in a sensitive way is a popular choice. Human employees (either paid or unpaid are used to annotate a series of photos in this example. Such annotated photos have been utilized to train models of supervised machine learning to properly categorize fresh pictures (Ofli et al., 2016).

2.2. FILTERING THE IMAGE STREAM FROM SOCIAL MEDIA

The social media picture data stream comprises a substantial volume of useless or duplicated material, posing a challenge for human annotations and computer processing. Frequently, individuals just retweet an old tweet. This "re-tweet" activity conveys essential information about the social agreement, but it provides no extra information from the standpoint of visual interpretation (Chen et al., 2013). On either side, some social media consumers upload unnecessary photographs, advertisements, or even pornographic material utilizing event-specific hashtags to market their stuff. Furthermore, neither the time nor the desire of human annotators is free or limitless. Each crowdsourced deployment entails a cost on the volunteer base and budget of humanitarian groups. Annotators can burn out (e.g., become less efficient owing to a loss of desire, fatigue, or stress) or quit entirely. It will impact the volunteer base of humanitarian groups. Because human annotations get a direct influence on the effectiveness of machine learning methods, defects in the comments may rapidly translate into flaws in the automatic categorization systems that are built. Thus, this is the highest need to get a large number of volunteers to offer annotations (such as quantity) and procedures to maintain the top quality of the annotations (Palen et al., 2010).

One approach to do this is to reduce the effort of human annotators by removing irrelevant text. We create the processing pipeline of the image depending upon perceptual hashing and deep learning for this aim, which may automatically (Peters and Joao, 2015):

- recognize and filter out photographs that aren't relevant or don't provide enough data for management and crisis response.
- remove duplicate or semi-duplicate photos that give no further information to categorization algorithms or humanitarian groups.

As a result, such filtration modules would assist annotators in concentrating their effort and time on just making sense of valuable picture material (Nguyen et al., 2017). They would also assist in decreasing the demand for computing sources as well as further enhancing machine categorization efficiency.

2.2.1. Relevancy Filtering

The idea of relevance is heavily influenced by the task's environment and needs. Modeling relevance is a difficult challenge since the context of relevance varies among disasters, and humanitarian groups, including within a long-running incident (such as conflicts, wars, etc.). Photographs of destroyed structures, for instance, can pique the attention of one humanitarian group, whereas pictures of injured individuals would pique the enthusiasm of another (Thom and Daly, 2016). What is judged unimportant, on the other hand, appears to be constant throughout crises and among numerous humanitarian groups. Pictures like celebrities, cartoons, banners, commercials, and so on, for instance, are all instances of irrelevant material (shown in Figure 2.1), and thus useless for disaster management and response. As a result, we present a relevance filtering strategy in this paper that emphasizes removing these irrelevant photos from our data processing pipeline (Sun et al., 2009).

Figure 2.1. Images from our databases of cartoons, banners, advertisements, celebrities, as well as other irrelevant images.

Source: https://www.slideshare.net/mimran15/image4act-online-social-media-image-processing-for-disaster-response.

2.2.2. De-duplication Filtering

Duplicate or near-duplicate photographs make up a major amount of the social media imaging information shared while disasters. For instance, there have been times when users just retweet a picture that has already been tweeted, or they share photos with minor changes, such as background adding, resizing or cropping, inserting text, altering intensity, and so on (shown in Figure 2.2). As a result of this posting behavior, a large number of near-duplicate photographs are produced, none of which offer considerable value to the collection of online data. As a result, we developed a de-duplication filtering approach in this study to eliminate redundant photos from the information processing pipeline (Alam et al., 2017).

Figure 2.2. Images that are near-duplicates from our databases.

Source: https://www.researchgate.net/figure/Examples-of-near-duplicate-images-found-in-our-datasets_Fig1_315786191.

2.3. ACTIONABLE INFORMATION EXTRACTION: DAMAGE ASSESSMENT

Since the imagery data of social media are cleaned from the redundant or irrelevant content and higher-quality annotations have been acquired from human annotators, machine learning models may be trained to analyze the subsequent collection of clean pictures to extract implementable data for different situational awareness jobs, including infrastructure damage evaluation, monitoring rescue operations, wounded person detection, and so on. This study focuses on the usage scenario of automatic severity evaluation of damage from photographs during natural disasters, one of the numerous potential usage cases for the suggested image processing pipeline upon social networks ((Rudra et al., 2016).

The primary situational consciousness needs of humanitarian response groups at the start of a disaster is assessing the amount of damage to determine the severity of devastation and organize relief activities appropriately. First responders must understand what sort of damage occurred and where it occurred. The present study on the usage of Twitter for damage assessment during an emergency focuses mostly on the content of the linguistic tweet (Cresci et al., 2015). With the current advancements in the computer vision field, particularly picture categorization, the majority of existing works for emergency preparedness don't yet utilize image information. We have just used cutting-edge computer vision techniques to conduct comprehensive damage assessment experiments utilizing photos from four main natural catastrophes (Nguyen et al., 2017).

Our objective is to discover whether or not this is feasible to judge the severity of the damage using photographs. For the evaluation of damage, we consider three levels: negligible harm, moderate damage, and severe damage (the Collection of Data and Annotation section contains further information on every damage level). Considering that thousands of photographs are broadcast on the platforms of social media like Twitter during catastrophes, a simple automatic damage evaluation system to determine if an image has damage or not would not be regarded as particularly useful for emergency workers (Zauner, 2010). Instead, we require a system that can recognize damage-related photos as well as estimate the severity of the damage (such as negligible, mild, and severe). This would substantially assist emergency responders in prioritizing and planning for the most serious situations initially. To our knowledge, no previous study has tackled the problem of determining the severity of damage (such as negligible, mild, and severe) from social media photographs in the way that ours does (Song et al., 2016).

2.4. STUDIES ON IMAGE PROCESSING

Entropy-based techniques, perceptual hashing, a bag of Words, and deep properties are all used in current duplicate picture identification research (An et al., 2017). Several such research employs hamming distance to calculate the resemblance of two pictures upon extraction of features. This necessitates the establishment of a threshold for detecting duplicate and near-duplicate photos. But, there is little research on how to describe this limit in the literature. We investigated deep features and perceptual hashing as well as identifying a viable way to determine this criterion in this work. Apart from one specific research that concentrated on getting visually irrelevant and relevant tweets having pictures (Chen et al., 2013), the majority of job on identifying relevant pictures is from the data retrieval domain, such as identifying images depending upon query. Unlike previous work, we describe relevance as "what is genuinely unrelated to the crisis occurrence depending upon the semantics of the circumstance at hand (e.g., severity evaluation of damage)."

There has been little research on leveraging social media information to develop classifiers for risk evaluation. Such research depends mostly upon the usage of standard Bag-of-Words characteristics in conjunction with classic machine learning classifiers like SVMs (Lagerstrom et al., 2016), and so produces limited results. Current breakthroughs in convolutional neural networks (CNNs), on either hand, have enabled even more performance increases. As a result, we perform comprehensive testing of domain adaptation and transfer learning algorithms for risk evaluation and disaster response using CNNs. An additional drawback is that there has been presented no publicly available damage evaluation dataset. By keeping our information available to the public, we want to shed light on this constraint (An et al., 2017).

As a result, we may describe our work's primary accomplishments as follows (Philbin et al., 2008):

i. We present strategies for eliminating duplicate, near-duplicate, and unnecessary picture material from chaotic social media imagery information.

ii. On the datasets of real-world crises, we demonstrate that deep learning models of the state of the art computer vision may be effectively modified to picture relevance and damage type categorization challenges.

iii. We illustrate that a large portion of the real-world crisis information gathered from online social media networks contains useless material using the proposed procedures.

iv. Our thorough experimental findings highlight the significance of the suggested picture filtering techniques for effective human and machine resource use. Our conclusions suggest that the purification of social media picture content allows humanitarian organizations to make better usage of their restricted human annotation budget throughout a crisis and increases the resilience and the machine learning models' quality outputs.

v. Ultimately, we show a comprehensive system that includes a real-time image processing pipeline for assessing social media information at the start of any incident. To illustrate the adaptability of the suggested social media image processing pipeline, we tested the system's efficiency in words of latency and throughput (Ke et al., 2010).

The following is how the rest of the article is structured. In the relevant Work section, we give a literature evaluation. In the collection of data and Annotation part, we go over our real-world disaster datasets and the procedure of labeling them. The Real-time Social Media Image Processing Pipeline part then introduces our automatic image processing pipeline, and the Experimental Framework section elaborates on our trials and outcomes. In the System efficiency Experiments part, we analyze the recommended system's efficiency in words of latency and throughput (Sukthankar et al., 2004).

There have been just a few studies assessing the social media picture content disseminated during crises, even though the abundance of text-dependent analysis of social media information for crisis management and response. One significant effort (Imran et al., 2013) offers a system termed AIDR, that primarily concentrates on analyzing and collecting tweets to ease humanitarian organizations' requirements. The AIDR method uses human volunteers to train robots to analyze tweets in real-time. Imran et al. provide a comprehensive survey, focusing mostly on text processing systems. Because this paper is about picture categorization and filtration, we'll go through the most up-to-date methods for duplicate image identification, image categorization, and image damage evaluation particularly (Imran et al. (2015).

2.4.1. Importance of Social Media Image Analysis

Peters and Joao have emphasized the usefulness of social media photos for disaster management (Peters and Joao 2015). The authors looked at messages and tweets from Instagram and Flickr during the 2013 floods in Saxony and discovered that photographs inside on-topic communications had been more related to the crisis, and the content of the image also gave crucial data about the disaster. Another work by Daly and Thom focuses on identifying photos retrieved from social media, such as Flicker, and analyzing whether a fire event happened at a certain location and time (Thom and Daly, 2016). Their research looked at the photos' spatiotemporal meta-data and found that geotags helped locate the fire-affected locations. Chen et al. investigated the relationship between photos and tweets, as well as their usage in categorizing visually irrelevant and relevant tweets (Chen et al., 2013). They created classifiers that included text, picture, and socially important contextual data (e.g., follower ratio, posting time, retweets, and number of comments) and achieved an F1 score of 70.5% in a binary categorization challenge, that is 5.7% greater as compared to the text-only categorization.

2.4.2. The Detection of Duplicate Image

There have been a range of methods for detecting duplicate and near-duplicate images, including MD5, locality sensitive hashing, perceptual hashing indexing dependent upon consistent randomized trees (Qiu et al., 2014).

The fingerprint of a picture obtained from multiple aspects of its content is represented by the perceptual hashing-based technique particularly. Histogram equalization or compression, resizing, and cropping may all change the digital representation of a picture. Traditional cryptographic hashing approaches are ineffective in capturing changes in binary representation, that are prevalent in near-duplicate pictures (Rivest, 1992). Perceptual hash functions, on either hand, retain picture perceptual equality and are thus capable of detecting even minor differences in the binary representation of 2 comparable pictures. Perceptual hashing's effectiveness for picture similarity measures is also documented in (Gong et al., 2015), wherein they demonstrate that perceptual hashing scales effectively for clustering a huge amount of pictures.

Duplicate image identification is aided by extracting deep characteristics (e.g., features acquired from deep network architecture) and then measuring the distance between the resultant deep picture properties in a deep learning-based technique. Zheng et al. demonstrated that deep features retrieved utilizing adversarial neural networks were effective in recognizing near-duplicate photos (Zheng et al., 2016). An et al. currently introduced a deep learning-based attribute extraction method accompanied by subspace dependent hashing (An et al., 2017). Computing resources and further enhancing machine categorization efficiency.

2.4.3. General Image Categorization

Studies on the categorization of an image at the cutting edge span from classifying photos and recognizing objects (Zhang et al., 2016) to creating captions (Xu et al., 2015). On large labeled picture datasets like ImageNet (Russakovsky et al., 2015) or PASCAL VOC (Van Gool et al., 2010) the majority of such experiments use distinct convolutional neural network (CNN) designs. GoogLeNet (Szegedy et al., 2015), AlexNet (Sutskever et al., 2012), and VGG (Zisserman and Simonyan, 2014) have been the most prominent CNN designs. The VGG is built on architecture with quite tiny (3 × 3) convolution filters and 16 to 19 levels of depth. VGG-16 is the name given to the 16-layer network. AlexNet is made up of five convolutional layers and three fully linked layers. GoogLeNet's architecture is made up of 22 pooling and convolutional layers piled on top of one another. The ImageNet Large Scale Visual Recognition Challenge (ILSVRC) in 2016, and ensembling techniques based on established CNN architectures like Inception Networks (Vanhoucke et al., 2015), Broad Residual Networks (Komodakis and Zagoruyko, 2016), and Residual Networks (He et al., 2016) revealed the best efficiency on image categorization task as 2.99% top five categorization error.

To get a competitive outcome, training, and developing the novel network architecture from scratch has been a technologically costly operation which also necessitates thousands of photos. Consequently, the common strategy is to adopt a pre-trained network to a target job that was trained on a larger dataset, such as ImageNet, which has a million pictures with one thousand classifications. This method is known as transfer learning in the literature (Girshick et al., 2014). The concept of transfer learning can be used in any of the following scenarios in the picture categorization task: Utilize the pre-trained model as a properties extractor, or utilize the network containing pre-trained weights it using the data from the current job. Several follow-up experiments have demonstrated that the characteristics learned automatically through such deep neural networks may be transferred to a variety of target domains (Donahue et al., 2014; Laptev and Sivic, 2014). It is particularly beneficial for training a large network to avoid overfitting whenever the target dataset is substantially small as compared to the basic dataset, for instance, but still obtaining state of the art efficiency in the target range (Pesaresi et al., 2007)Image categorization for damage evaluation

The usage of computer vision algorithms for assessing damage from pictures has not been thoroughly investigated in the literature. Few research findings in other ranges of research (e.g., remote sensing) have evaluated the extent of damage from aerial and satellite pictures gathered from disaster-hit areas. The most relevant and most current research depending on social media information includes (Lagerstrom et al., 2016) and both of which evaluate social media photographs in a binary categorization context for fire or not-fire detection scenarios. We are currently working on a more difficult challenge because we are not limiting the task to a certain catastrophe type or a binary categorization setup (Thom and Daly, 2016).

Many of the previous studies used a bag of visual words approach to damage evaluation, combining handcrafted characteristics like Histogram of Oriented Gradients (HOG) and Scale Invariant Feature Transform (SIFT) having classical machine learning classifiers like Random Forest (RF) and Support Vector Machines (SVMs) (Schmid et al., 2006). SIFT and HOG are two features that recognize and characterize local data in pictures. The bag of visual words method then uses vector quantization to turn such picture attributes into visual words. We investigated the transfer learning strategy utilizing CNN for crisis management and response in the study because there have been few studies on risk evaluation from photos and they have been primarily confined to the usage of the bag of visual words method (Dance et al., 2004).

Table 2.1. Details About Each of the Four Disasters, Including the Year and Number of Photographs

Year	Disaster	Number of pictures
2016	Ecuador Earthquake	65,000
2016	Hurricane Matthew	61,000
2015	Nepal Earthquake	57,000
2014	Typhoon Ruby	7,000

2.5. DATA COLLECTION AND ANNOTATION

During the aftermath of four main natural catastrophes such as Typhoon Ruby, the earthquake that struck Nepal, the earthquake that struck Ecuador, and Hurricane Matthew, we collected images from social media platforms using a system called AIDR (Imran et al., 2014; Kuminski et al., 2014), which is accessible to the public. The gathering of data had been based on hashtags and phrases that were particular to the event. Twitter had been selected for the data gathering rather than other social media sites like Flickr and Instagram for a variety of reasons (Alam et al., 2018). These reasons include: Firstly, Twitter makes it easier and more effective to automatically gather data by providing a method through its application programming interfaces. Secondly, photos that are connected to tweets have textual material as well, which may be utilized in a multimodal analysis strategy if one so chooses. The total number of photos that were initially gathered for each dataset is presented in Table 2.1. Figure 2.3 displays some representative photos taken from each of these datasets (Feng et al., 2014).

Figure 2.3. Images from several disaster datasets with varying levels of devastation.

Source: https://www.researchgate.net/figure/Sample-images-from-SAD-image-dataset-left-right-a-fire-disaster-b-flood-disaster-c_Fig2_343262615.

2.5.1. Human Annotations

For the training objective and assessing machine learning models for picture filtration and categorization, we obtained human labels. While photographs from Twitter had been utilized in other projects, we concentrated on damage severity evaluation because this is among the most important situational awareness jobs for several humanitarian organizations. We acquired human annotations in two distinct situations for such purpose (Turker and San, 2004).

The very first set of annotations including all four catastrophes had been acquired from AIDR with the help of volunteers from the Stand by Task Force (SBTF) community. Such volunteers had been called to help amidst a crisis (Aroyo and Welty, 2015). Volunteers were used to annotate photographs while Hurricane Matthew tweets had been gathered from Twitter. This procedure entailed gathering tweets, determining if they had an image URL, obtaining real photos from the Web if they did, and then allocating them to annotators. It is impossible to predict the number of photographs to be tagged in such real-time crowdsourced settings. Controlling duplicate photographs is also a difficult challenge (Attari et al., 2017). As a result, our volunteers had been exposed to duplicate cases as well. To annotate photographs in the 2nd option, we utilized Crowdflower, a paid crowdsourcing tool. From the Ecuador Earthquake, Nepal Earthquake, and Typhoon Ruby databases, a random selection of one thousand photos was chosen. Every image had to be tagged by at least three human annotators in both settings to ensure high-quality annotations, and a final label was chosen utilizing the main vote approach. Images containing less than three annotations had been deemed untrustworthy and were thus eliminated (Kerle et al., 2015) (Figure 2.4).

Figure 2.4. Images from several disaster datasets with varying numbers of damage.

Source: https://www.researchgate.net/figure/Sample-images-with-different-damage-levels-from-different-disaster-datasets_Fig3_322138548.

2.5.2. Annotation Instructions

In this job, we decided to use a three-level graphic to depict the severity of the damage. Having a 3-level severity evaluation technique rather than a finer scale rating scheme (e.g., 0–8) allows humanitarian responders to swiftly examine photographs indicating just serious damage, rather than having to go into several image buckets reflecting various scales like 6, 7, and 8. Furthermore, whenever a finer scale grading scheme has been anticipated, acquiring consistent and accurate ground truth via human annotators for the degree of damage appears to be a considerably more difficult process (Malik and Darrell, 2014).

The crowdsourced work summary, damage types, and terminology are all shown here. Annotators had been instructed to follow the guidelines and annotate photographs using category definitions (Afzal and Robinson, 2011).

Description of the job: The goal of this exercise is to determine the extent of damage in a picture. The degree of physical devastation depicted in a photograph determines the severity of the damage. Only physical damages such as damaged bridges, fallen or broken buildings, wrecked or cracked roadways, and so on are of concern to us. Signs of smoke from a fire on a structure or bridge are an instance of non-physical damage; nevertheless, we don't include these damage kinds in this work (McNitt-Gray et al., 2007).

Severe Damage: Images that depict the complete devastation of infrastructure are classified as severe damage. A badly damaged infrastructure might include a non-livable or useful building, a non-drivable road, or a non-crossable bridge (Speck et al., 2011).

Mild Damage: Damage that is worse than mild [damage], having up to 50% of a structure in the focus of the photograph, for instance, suffering a partial loss of roof/amenity. Only a portion of the structure may need to be shut down, while the rest can be utilized. If a bridge may still be utilized, but parts of it are damaged and/or require repair, the bridge is still useful. Furthermore, if the road is still accessible, part of it must be shut off due to damage to the situation of a road picture. Such damage must be far worse than what we observe from normal wear and tear. Images that depict infrastructure with little to no damage (save for tear and wear due to age or degradation) fall into this category (Rahman et al., 2019) The combined human annotation findings both from annotation settings are shown in Table

2.2. Images with 3 or greater three annotations from various annotators had been deemed trustworthy and utilized in our tests; alternatively, they had been rejected and not included in Table 2.2. Only 336 photos had been identified by three or more human annotators for the Hurricane Matthew collection. Furthermore, because the annotation procedure had been carried out on the raw picture sets, with no pre-filtering to clean the datasets, the labeled datasets produced have duplicate and unrelated pictures. Generally, we collected a substantially higher number of tagged photos for catastrophes like the Typhoon Ruby (equivalent to 7,000) and Nepal Earthquake (equivalent to 25,500) than we did for the Hurricane Matthew (equivalent to 350) and Ecuador Earthquake (equivalent to 2,000) (Rashtchian et al., 2010).

Table 2.2. The Number of Classified Photos in Every Damage Category for Every Dataset

Category	Ecuador Earth-quake	Nepal Earth-quake	Hurricane Earthquake	Typhoon Earth-quake	Total
None	946	14,239	132	6,387	21,704
Mild	89	2,257	94	338	2,778
Severe	955	8,927	110	88	10,080
Total	1,990	25,423	336	6,813	34,562

2.6. IMAGE PROCESSING PIPELINE FOR SOCIAL MEDIA IN REAL-TIME

Humanitarian groups need real-time perceptions from the information shared on social media at the start of an emergency occurrence to be effective during catastrophes. To meet such time-sensitive information requirements, data must be processed as it appears (Maynard et al., 2017). That implies the system must take information from online platforms since it has been posted, analyze it, and analyze it in real-time to get insights. We currently demonstrated an autonomous image processing pipeline to accomplish such capabilities (Imran, Alam, & Ofli, 2017). The pipeline is depicted in Figure 2.5 along with its numerous modules and key elements, which we will discuss later (Melenli and Topkaya, 2020).

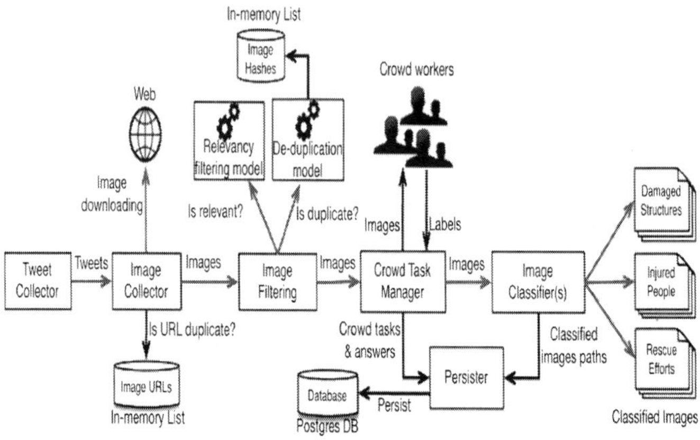

Figure 2.5. Automatic image processing pipeline.

Source:https://www.researchgate.net/figure/Automatic-image-processing-pipeline_Fig1_322138548.

2.6.1. Tweet Collector

The collection of real-time tweets from the Twitter streaming API comes under the jurisdiction of the Tweet Collector module. The system has the capability of being configured to gather tweets from various catastrophes that are occurring simultaneously (Chen and Freire, 2020) (Figure 2.6).

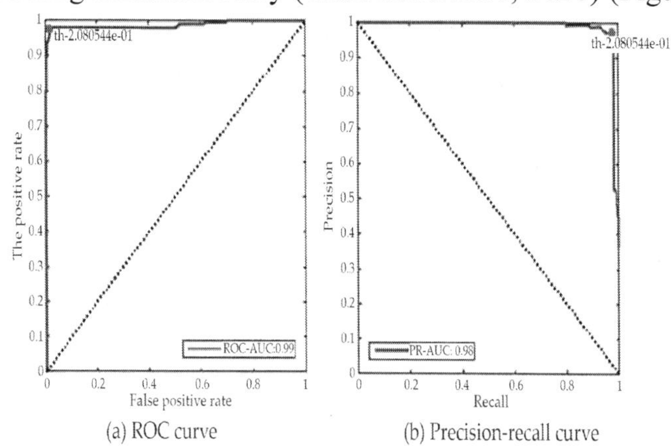

Figure 2.6. Calculation of the distance threshold d for detecting duplicate images utilizing a technique based on deep learning.

Source:https://journalofbigdata.springeropen.com/articles/10.1186/s40537–019–0197–0.

The user provides hashtags, keywords, geographical bounding areas, and Twitter users to establish a catalogue in the system for a certain event (e.g., an earthquake). Just geo-tagged tweets have been gathered when using the geographical bounding box option, however, you may utilize both the bounding box choices and keywords to receive tweets matching either of the tweets and keywords inside the designated geographical boundaries. Even though the pipeline may be expanded to ingest photographs from other platforms of social media like Instagram, Facebook, and others, we will fully concentrate on images published through the Twitter platform (Quirin et al., 2018).

2.6.2. Image Collector

The Image Collector module consumes tweets gathered by the Tweet Collector to obtain image URLs from the gathered tweets. Following that, the Image Collector uses the retrieved URLs to download photos from the Internet (e.g., in certain cases from Instagram or Flickr). Owing to retweets, a high percentage of the tweets gathered include duplicate URLs (Yanai, 2007). To prevent downloading the duplicate photographs, the system keeps a list (e.g., an associated hash map) of distinct URLs in memory (e.g., the Image URLs information in Figure 2.4). Each of the system's collections has its information. To discover duplicates, a freshly incoming picture URL is 1st verified against the in-memory database. If the URL is distinctive, this is added to the list of Image URLs and pushed into a separate queue that maintains track of all the URLs that are waiting for their respective pictures to be retrieved from the Web. If a URL has been discovered to be duplicated, this is simply ignored. The time complexity of adding and finding a member in the in-memory list is O(1), while the space complexity is O(n) (Basbug, 2020).

Once the system has a queue of image URLs, it begins downloading photos, which are subsequently published to a set specific Redis channel. All of that channel's subscribers receive photographs as soon as they have been downloaded (Itaki et al., 2018).

2.6.3. The Filtering of Image

The goal of such a module is to reduce the amount of noise in the incoming imaging data stream. As previously stated, in this study, two categories of photos are deemed noise: (i) duplicate or near-duplicate images, and (ii) photographs that have been unrelated to the response to the disaster (Zhang and Gunturk, 2008). The Image Filtering module accomplishes two major tasks by reducing noisy stuff from the picture data stream: Firstly, to filter out extraneous photographs provided during catastrophes to present disaster management with only relevant and helpful material. Secondly, deleting unnecessary and near-duplicate photos, as indicated in the pipeline (e.g., the module of Crowd Task Manager), improves the performance of human employees (Ding et al., 2011). If human employees who have been intended to label photos for different machine learning jobs have been kept excessively occupied labeling several duplicate or irrelevant images, their time and hence the money would be lost. As a result, the module of image filtering in Figure 2.4 has two submodules: relevance filtration of image and de-duplication filtering of image, which will be discussed later. The module of image filtering sends the picture byte stream to such submodules, which aggregate its outputs into a JSON7 stream that the Crowd Task Manager module may use (Lee, 1981).

2.6.3.1. Relevancy Filtering

The relevance filtration submodule receives an image byte stream and processes it via a relevancy classifier. This process results in the generation of a confidence score as well as a class label. Among the most cutting-edge CNN architectures, known as VGG-16, is utilized to determine whether or not the picture that has been presented is significant (Singh et al., 2012). In the section titled "Relevancy Filtering Experiments," you will find a wealth of additional information on the construction and operation of the relevance classifier. The relevancy filtration submodule sends its categorization findings to the image filtration module, which then processes those findings (Meng et al., 2009).

2.6.3.2. De-duplication Filtering

The de-duplication filtering submodule takes an image byte stream as input and uses the perceptual hashing technique to evaluate if a particular picture is an identical or near-duplicate of earlier seen pictures, comparable to the relevancy filtering submodule (Hamming, 1950). To be more explicit, a hash value is produced for every incoming picture and contrasted to the in-memory image hashes to identify near-duplicate or duplicate situations depending upon their Hamming distances. The use of this strategy is motivated by the fact that this is computationally easier and performs quite well when compared to the deep learning-based method. We have undertaken studies utilizing deep features (Onno and Guillemot, 1993) (Figure 2.7).

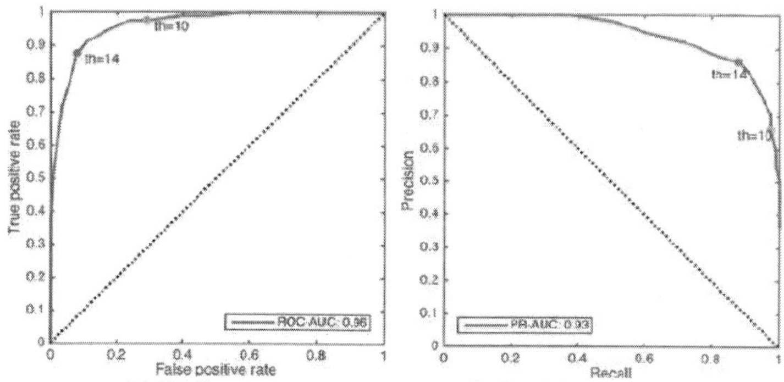

(a) ROC curve (b) Precision-recall curve

Figure 2.7. Using a perceptual hashing-based technique, we estimate the distance threshold D for duplicate picture identification.

Source: https://www.researchgate.net/figure/Estimation-of-distance-threshold-d-for-duplicate-image-detection_Fig4_315883693.

2.6.4. Crowd Task Manager

The Standby Task Force (SBTF) volunteers are assigned picture tagging jobs via the Crowd Task Manager module. A task, also known as a classifier (more on classifiers in the following part), is created by an end-user and contains a set of classifications (such as little-to-no damage, mild damage, and severe damage). A human labeler is shown a picture and a list of classifications by the Crowd Task Manager. The labeler assigns a suitable label to the image, that is subsequently used as a training instance (Labani et al., 2018).

2.6.5. Image Classifiers

After deleting irrelevant and duplicate photos, the system enables end-users (such as crisis managers) to construct photograph classifiers tailored to their unique information requirements. A damage evaluation classifier, for example, maybe constructed to collect photos that reveal a certain form of damage (Delpueyo et al., 2013). Likewise, a wounded people detection classifier may be built to collect all photographs of people having injuries. It is also possible to design many classifiers to categorize photos concurrently (e.g., both damage evaluation and injured people detection classifiers). Two (binary) or more classes may make up a classifier (multi-class). Eventually, in the Experimental Framework part, further information on training photograph classifiers for various utilization cases is offered. The Image Classifiers module gets two kinds of photos from the Crowd Task Manager for two purposes, as illustrated in Figure 2.5. Human-labeled pictures are one type of image that is utilized to train user-defined image classifiers. Unlabeled photos, on either hand, are automatically classified by the computer utilizing the classifiers it has created. The module of Image Classifiers then aggregates the classified data (confidence ratings and class labels) and sends it to the Persister module to be stored (Auffarth et al., 2010).

2.6.6. Persister

All database-specific actions, like insertion of picture storage, meta-data, and retrieval of classifier predictions, are handled by the Persister module. Additionally, it saves machine-tagged photos to the file system. To save tweets and their metadata, we utilize the PostgreSQL9 database (Bozdag, 2013).

Utilizing Redis channels, all of the modules listed above communicate with one another (e.g., data flow). In addition, each module has its collection of RESTFul APIs, which make it possible for the module to communicate with the outside world (for instance, through UI interactions) and, if required, adjust the values of its parameters. In Figure 2.8, live streams that are transmitting data items are represented by the red arrows, while non-streaming communications are represented by the black arrows. To construct the system, the programming language known as Java Enterprise Edition (J2EE)10 is utilized (Kawata et al., 2015).

2.7. EXPERIMENTAL FRAMEWORK

The development and design of the de-duplication filter, relevance classifier, and damage evaluation classifier are discussed in this section. The efficiency of our picture filtering and damage evaluation modules is then evaluated in four distinct testing conditions (Al-Ayyoub et al., 2015). To measure the efficiency of various elements of the system, we employ well-known metrics like precision, accuracy, F1-score, recall, and AUC. The fraction of correct +ve and -ve predictions are used to calculate accuracy. The amount of real +ve predictions divided by the total number of +ve forecasts is known as precision. The number of real +ve predictions divided by the total number of +ve cases is known as recall. The harmonic mean of recall and accuracy is the F1 score. The region under the precision-recall curve is used to calculate the AUC (Frith et al., 1999).

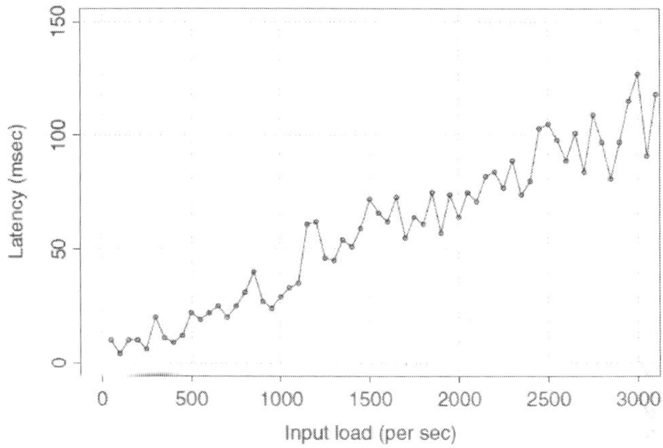

Figure 2.8. In the image collector module, there is the latency for URL de-duplication.

Source: https://www.researchgate.net/figure/Latency-for-the-URL-de-duplication-module_Fig2_322138548.

2.7.1. Relevancy Filtering Experiments

To decide if a picture is relevant, the Relevancy Filtering module uses a CNN. We utilize the DeepLearning4J toolkit to train a relevant model utilizing the VGG-1611 architecture (Simonyan and Zisserman, 2014). It's worth noting that the human annotation procedure described in the Annotation section and Data Collection had been created only to evaluate the extent of damage seen in a picture, with no consideration given to the image's actual content. As a result, we don't have any human annotations that may be used to evaluate the relevance of an image's content (Bin et al., 2017). To determine if a picture is relevant or not, one way will be to create a hand-design (a collection of characteristics) or a set of rules. Therefore, we decided not to utilize this method to avoid creating a skewed or limited definition of relevancy which can lead to the omission of potentially related and important material. Rather, we used the human-labeled data to develop a collection of picture characteristics that reflect the subset of irrelevant photos in our datasets, as detailed in the following sections (Lavrac et al., 1998).

Firstly, we assessed all of the photos that were originally labeled for the damage evaluation job (see the Collection of Data and Annotation section). Secondly, all photos in the mild and severe classifications were judged significant. Thirdly, we noted that the no class includes two kinds of photographs (Amasaki, 2019): (i) those had been associated with the disaster event and didn't simply depict any damage, and (ii) those had not been related to the disaster event at all, or for which the relationship might not be immediately understood based on the content of image alone. Fourthly, we determined that this final group of photographs in the no class must be deemed unimportant. We utilized the original VGG-16 model (Simonyan, & Zisserman, 2014,) to categorize every photograph into one of the ImageNet object classes to determine the set of photos under the no-class that may be declared irrelevant. The purpose of this study was to determine which ImageNet object classes occurred most frequently in the irrelevant section of non-category photos (Nelson et al., 2011). We examined the distribution of the most common ImageNet object classes for non-category photos for this study. Then, we examined the 50 most prevalent object classes (that also included half of the non-category photos) and retained only those that occurred relatively seldom (e.g., a minimum of ten times less frequently) across the relevant image collection. As a consequence of this study, we chose 14 ImageNet object classes (such as suit, website, envelope, lab coat, candle, dust jacket, vestment, menu, monitor, television, puzzle, street sign, screen, and cash machine) to detect irrelevant photos in our none category. Therefore, this method produced 3,518 irrelevant photos. We then utilized this dataset to train a binary relevance classifier (Peters et al., 2013).

When there is a small amount of data available in the new domain, the transfer learning approach is effective for performing visual identification tasks. As a consequence of this, we utilized a transfer learning method by using the current weights of the pre-trained VGG-16 network as initialization for the same network when we were fine-tuning it on our training dataset. In addition to this, we changed the last layer of the network so that it can deal with binary classification problems (e.g., two classes in the SoftMax layer) rather than the 1,000-class classification that it was designed for (Lu et al., 2010). As a result, we were able to move the network's properties and parameters from the wide domain (large-scale picture categorization) to the particular domain using this transfer learning technique (such as relevant-image categorization). During the fine-tuning of the network of VGG-16, we utilized 60% of our 7,036 photos for training and 20% for validation. The efficiency of the fine-tuned network was next assessed on the remaining 20% of the dataset (Gay et al., 2009).

The efficiency of the resultant relevance filter on the test set is shown in Table 2.3. Because the irrelevant and relevant photos in our training dataset have radically distinct visual properties and content, the binary classifier performs almost flawlessly (as may be shown in the example photos in Figures 2.1–2.3). This satisfies the original relevancy filtration goal of excluding those photos that have been unrelated to the work at hand. Remember that these 7,036 photos are just used for relevance filter modeling, while the remaining 27,526 images are used for the rest of the tests (Tseng et al., 2014).

2.7.2. De-duplication Filtering Experiments

The following is the method we used to recognize exact and near-duplicate photos. When the gap between two photos (a, b) is less than the threshold, a, and b are deemed near or exact duplicates. Various distance functions may be used based on the situation, and the threshold may be learnt empirically. We studied two ways to detect duplicate photos in this study: (i) a deep learning-dependent technique and (ii) a perceptual hashing-dependent technique (Koppula et al., 2010).

We looked at a perceptual hashing-based technique in the 1st experiment. The perceptual hashing approach collects particular attributes from every picture, calculates a hash value (e.g., a binary string of length 49) for every photograph based on such properties, and then contrasts the resultant hashes pair to determine how the images are identical. While in an event, the system keeps track of hashes calculated for a collection of distinct pictures received from the Image Collector module inside a list (e.g., an in-memory data structure) (Lu et al., 2011). To identify if a newly received picture is a duplicate of an already observed photo, the new photo's hash value has been generated and contrasted to a list of previously recorded hashes to estimate its distance from the present hashes. In this scenario, we compare two hashes using the Hamming distance. If the gap between a newly arriving image's hash and a hash in the list is less than (threshold) d, the newly received photo is considered a copy. The most recent 100K hashes are always kept in physical memory. This figure is dependent on the amount of RAM accessible in the machine (Ma et al., 2010).

Defining an adequate distance threshold d is one of the crucial factors to consider while attempting to determine duplicate photos. To achieve this goal, we performed manual inspections on all picture pairings that had a Hamming distance ranging from zero to twenty. Pairs with a distance 'd' greater than twenty appeared to be insignificantly dissimilar, and as a result, they had not been chosen for manual annotation. After this, we visually inspected every pair of remaining photos and were given a value of one of the photographs in that pair must be deemed duplicates, and a value of 0 if they shouldn't be. Because of this approach, we were able to gather 550 picture pairings, which resulted in a total of 1,100 photographs with ground truth annotations indicating whether or not they were duplicates (Xiai et al., 2016).

We retrieved deep properties from the fc7 layer of the VGG-16 network trained upon ImageNet (e.g., the final 4,096-dimensional layer before the soft-max layer) for all 550 picture pairings for the experiment of deep learning-based, and calculated the Euclidean distance between every image pair (Patil and Dongre, 2020).

Figure 2.5 shows the de-duplication tests' Receiver Operating Characteristic (ROC) and precision-recall curves. By examining the best operating point of the ROC curve presented in Figure 5a (indicated with a red dot), we derive the proper threshold value for our de-duplication filter as d 14. This method gives us the optimal balance between the cost of not detecting genuine positives and the expense of raising false alarms. As shown in Figure 5b, that threshold setting yields accuracy and recall score of nearly 0.9. Setting the threshold value to d 10 (shown with a green dot in Figure 2.5) produces a recall score of almost 1.0 and an accuracy score of 0.65, which is almost flawless recall (that is, practically no chance of missing any potentially unique picture) at the price of raising false alarms. The system also obtained AUC 0.96 for the ROC and AUC 0.93 for the precision-recall curve, as shown in Figure 2.5 (Jiang and Liu, 2012).

The ROC and precision-recall curves for the experiment wherein we substituted perceptual hash properties with deep features for de-duplication filtering are shown in Figure 2.6. By these graphs, d 0.208 is the right threshold value (e.g., an optimal operating point on the ROC curve) for our test set, achieving almost flawless scores in all metrics (AUC, recall, and precision) (Kudela et al., 2018).

We use our suggested relevance and de-duplication filtering modules to 27, 526 photos in our dataset to demonstrate the advantages of the image filtering elements and to identify what fraction of the material on online social networks is possibly relevant. The number of photos maintained in the dataset after every image filtering process is shown in Table 2.4. As predicted, relevant filtering removes 8,886 of the 18,186 photos in the none category, a reduction of about 50%. Although certain photos have been deleted from the severe and moderate categories (212 and 164, correspondingly), the error margins for the trained relevance classifier are within the appropriate range of two percent (Yang et al., 1996). The de-duplication filter, on either side, eliminates a significant percentage of photos from all classifications, namely 58%, 50%, and 30% from severe, mild, and none, correspondingly. The fact that social media consumers prefer to re-post the most relevant information quite often explains the comparatively greater removal rate for the severe and moderate classifications. As a result, our picture filtering process decreases the size of the raw image data gathering by nearly a factor of three (e.g., a 62% reduction) while keeping the most relevant and interesting image material for future analyzes (Tzeng et al., 2007).

Table 2.4. Number of images that remain in our dataset after each image filtering

Category	Raw Collection	After Relevancy Filtering	After De-duplication Filtering	Overall Reduction
Severe	7,501	7,289	3,084	59%
Mild	1,839	1,675	844	54%
None	18,186	9,300	6,553	64%
Total	27,526	18,264	10,481	**62%**

2.7.3. Experiments on Damage Evaluation Using Image Filtering

One of the primary objectives of our work was to comprehend the impact of picture filtering on the task-specific classifier that followed. As a result, we have undertaken tests to determine the impact of duplicate and irrelevant pictures on human calculation and machine learning. For the trials, we created four parameters for picture filtering: relevance and de-duplication filtering, accompanied by damage evaluation categorization (Aswal et al., 2022).

S1: We conduct experiments on unprocessed data by preserving irrelevant and duplicate photos. The findings from such a setting have been used as a baseline for the subsequent settings.

S2: The S1 dataset is refined by eliminating only duplicate photos. The objective of this setting is to understand the distinction between duplicates and unique items.

S3: Likewise, we just delete irrelevant photos from the S1 dataset and retain the remaining data. The objective of this study is to examine the impact of deleting irrelevant photos from the training set.

S4: We eliminate redundant and unnecessary pictures. This is the optimal configuration, which is also implemented in the pipeline we suggest. It is anticipated that this configuration would outperform others in terms of budget use and the efficiency of the machine (Kudela et al., 2015).

We used the same method of fine-tuning a pre-trained VGG-16 network to construct a damage evaluation classifier (such as similar to designing the relevancy filtering model). Therefore, when it came to training our damage evaluation model, we used a somewhat unique approach: (i) Instead of using a train/validation/test data split, a network is trained for 3-class classification using the classes of severe, mild, and none, and (ii) the performance of the resultant damage identification classifier is tested using 5-fold cross-validation. This ensures that the performance is as accurate as possible (Elmgerbi et al., 2021).

We decided to establish a budget of 6,000 USD to model the aforementioned four different environments. For the sake of simplicity, we will suppose that the cost of having one image identified by human labor is one dollar. In Stage 1, the system uses up its whole budget of 6,000 USD to obtain 6,000 tagged photos from the raw data. A significant number of these images seem to be duplicated. To emulate this, we choose 6,000 photos at random from the labeled data while keeping the original class distributions the same, as depicted in the S1 column of Table 2.3. After that, we train a damage identification classifier using these 6,000 photos as was explained earlier, and then we provide the efficiency of the classifier in the S1 column of Table 2.3 (Sulaiman et al., 2017).

In Stage 2, we take the same 6,000 photos from S1 and put them via our de-duplication filter to remove any duplicates, before training a damage identification classifier on the cleaned subset. The class-wise allocation of the remaining pictures after de-duplication is shown in the S2 column of Table 2.3. In the severe, mild, and none classifications, 598, 121, and 459 photos are tagged as duplicates and eliminated, correspondingly. This reveals a budget waste of 1,178 USD (0.20%) in S1 that might are avoided if the de-duplication approach had been used. The damage identification classifier's efficiency on cleaned data is displayed in the S2 column of Table 2.3 (Yuen et al., 2009).

In S3, we do a relevance filtering operation on the raw data before sampling 6,000 photos from the clean set of 18,264 pictures. It's worth noting that this 6,000-image specimen might potentially have duplicates or near-duplicates. Even though the training data for S1 and S3 aren't identical, we may still evaluate the damage identification model's efficiency without or with irrelevant pictures. Table 2.3 shows that the scores for the none category in S3 are less than those in S1, although the scores for the mild and severe categories in S3 are greater than those in S1. We can observe that the overall F1 score for S3 is one percent greater than the overall F1 score for S2, although the macro-averaged AUC values appear to be the same after macro-averaging the scores for all categories. We don't aim to express more on this comparison because having near-duplicate or duplicate photos in the training and test sets leads to untrustworthy model outputs (Ostachowicz et al., 2013).

In S4, we remove irrelevant and duplicate photographs from the raw data gathering, then choose six thousand images from the remaining 10,481 images that are clean. The findings of the damage identification experiment on the chosen dataset, that doesn't have irrelevant or duplicate photos, are presented in the S4 column of Table 2.3. When comparing the performance results for S3 and S4, we can see that deleting duplicate photos from the training data removes the spurious gain in performance scores, which is consistent with the trend seen between S1 and S2 (Bottou et al., 2013).

2.8. SYSTEM PERFORMANCE EXPERIMENTS

We used simulations to do thorough stress testing on the planned system to better comprehend its scalability. The tests were carried out on an iMac have a 32GB RAM configuration and a 3.5 GHz quad-core CPU. The simulation trials had been carried out with,28K pictures. To simulate the behavior of the Tweet Collector module, we created a simulator software. The simulator may be configured to publish variable numbers of photos (such as batches) to Redis channels in a particular unit of time, such as 50 or 1,000 images per second. We gradually raised the input load (e.g., the number of photos) while retaining a unit time to examine a module's efficiency in words of latency and throughput (e.g., in our case 1 second). The time it takes a module to process an image is known as latency. Because we're dealing with a batch of photos, the latency is calculated as the total of a module's response time for all of the images in the batch, which may then be divided by the collection of pictures to produce an average latency per image. The throughput, on either side, is calculated as the number of photos processed per unit time by a module (Wang et al., 2012).

Table 2.3. Recall, Precision, AUC, and F1 Scores: S1 (with Duplicates + with Irrelevant), S2 (Without Duplicates + with Irrelevant), S3 (with Duplicates + Without Irrelevant), S4 (Without Duplicates + Without Irrelevant)

	S1				S2				S3				S4			
	AUC	Pre.	Rec.	F1	AUC	Pre.	Rec.	F1	AUC	Pre.	Rec.	F1	AUC	Pre.	Rec.	F1
None	0.98	0.91	0.96	0.94	0.98	0.91	0.97	0.94	0.94	0.86	0.93	0.90	0.95	0.86	0.95	0.91
Mild	0.31	0.48	0.18	0.25	0.26	0.41	0.12	0.18	0.37	0.53	0.20	0.29	0.30	0.55	0.14	0.23
Severe	0.95	0.88	0.89	0.88	0.91	0.85	0.84	0.84	0.95	0.88	0.91	0.90	0.91	0.86	0.85	0.86
Average	0.75	0.74	0.68	0.69	0.72	0.72	0.64	0.65	0.75	0.76	0.68	0.70	0.72	0.76	0.65	0.67

2.8.1. Image Collector Module

The Image Collector module keeps track of the entire different image URLs. One technological problem is determining the ideal size of the URL list to avoid increasing system latency during the search operation. We utilize the simulator to inject tweets containing picture URLs into the Image Collector module having various input loads to find the best size. The delay for URL de-duplication in the Image Collector module is shown in Figure 2.7. As previously mentioned in the Image Collector subsection, we keep an in-memory list (e.g., an associated hash map data structure) for URL de-duplication, that has a constant temporal complexity of O for adding and finding an element. Figure 2.7 shows how the system efficiently removes duplicate URLs while keeping a low latency. As the input load grows, the delay increases somewhat. From a performance standpoint, we can see that the Image Collector module can add and search URLs in a linear unit of time. Even though the system is capable of handling heavy input loads, we have established a restriction of 100K for the distinct picture URL list. As a result, whenever the in-memory list hits this limit, the oldest URLs have been eliminated (Burger et al., 2000).

2.8.2. De-duplication Filtering Module

Following the download of photos, the next step is to filter out duplicates, as described in the Image Filtering sub-section. Extracting image hashes, saving them in an in-memory list (e.g., an associated hash map data structure), and evaluating if a new incoming image hash matches with existing hashes depending upon a distance threshold are all part of the image de-duplication method. The results of our simulation trials for the de-duplication filtering module are shown in Figure 2.8. We see that when the batch size grows larger, the latency changes, but the throughput remains rather constant. This may process about eight photos per second on average. We limited the size of the in-memory list to 100K for this module. When the in-memory list hits its limit, we eliminate the oldest picture hashes in the same way we do with URL duplication (Krallinger et al., 2015).

2.8.3. Relevancy Filtering Module

Figure 2.9 shows the results of the Relevancy Filtration module simulation trials in terms of throughput and latency. The computational cost of our relevance categorization model is high. Figure 2.9 shows that it may only analyze nine photos per second at maximum, with delay varying based on input load. Furthermore, if the stress testing tests are done on a GPU-based system instead of a CPU-based system, as in this situation, one may expect to see a large increase in throughput and latency (Clune et al., 2014).

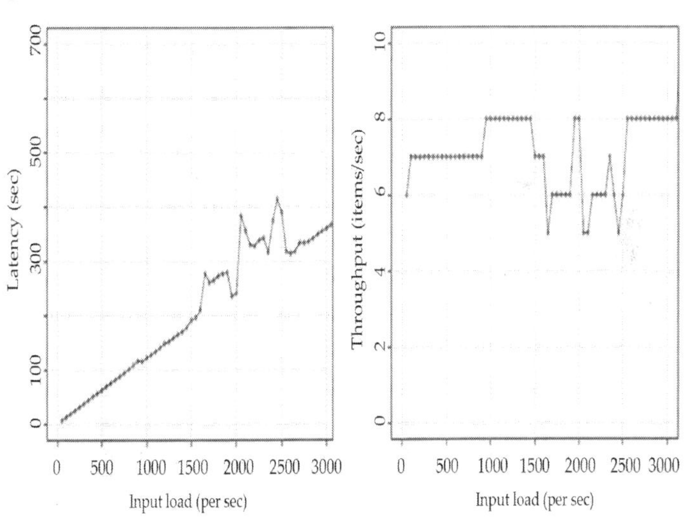

Figure 2.9. Throughput (right) and latency (left) for the de-duplication filtering module.

Source: https://www.mdpi.com/1424–8220/18/11/3832/htm.

According to the results of a comparison with a benchmark, the classifier is capable of assigning a label to around ten photos every second (Wang, Shi, Chu, & Xu, 2016).

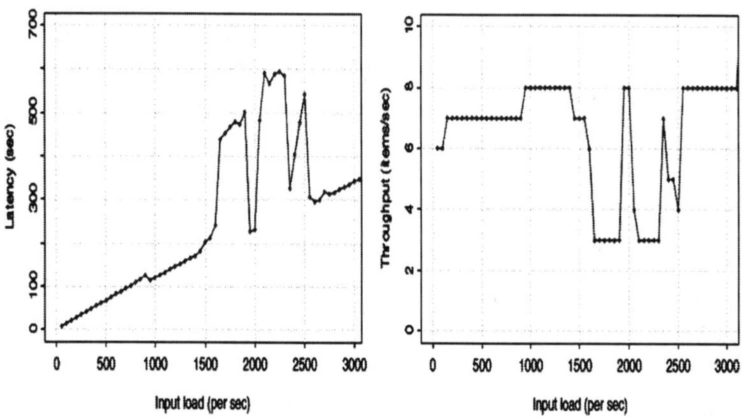

Figure 2.10. Latency (left) and throughput (right) for the relevancy filtering module.

Source: https://www.semanticscholar.org/paper/Joint-Optimization-of-Latency-and-Deployment-Cost-Wang-Ji/f31d5956b660fc690e286ff22c7a039da2d-8fec8.

Our exploratory investigation revealed a comparable processing time. A GPU-based system is projected to have considerably superior latency and throughput efficiency than the existing CPU-based system, depending on the same logic as the Relevancy Filtering module trials (Petrillo& Baycroft, 2010).

2.8.4. Latency of the Overall System

Even while certain portions of the system are dependent on external variables like local network infrastructure efficiency, picture download speed from the Web, etc, this is useful to look at the latency of the entire image processing pipeline. As a result, we used a local server to host pictures in our simulations. This delay would rise in a real application when the system downloads photos from external services. We discovered that as the batch size is increased, the latency increases dramatically. We may deduce that the entire system may handle fifty photos per minute on average (Aytar et al., 2016).

2.9. SUMMERY

When disasters strike, user-generated information on social media may help with crisis management and response. Analyzing this high-velocity, data high-volume, on either hand, is a difficult issue for humanitarian groups. Existing research points to the use of imaging data shared on social media during catastrophes. Furthermore, because there are so many redundant or irrelevant photos, effective use of the imaging material, whether through crowdsourcing or machine learning, is difficult. We provided a social media image processing pipeline in this research, which included two kinds of noise filtering (image de-duplication and relevancy) as well as a damage identification classifier. We employed a transfer learning strategy depending upon state-of-the-art deep neural networks to filter away irrelevant visual material. We used perceptual hashing algorithms for picture de-duplication. We have conducted thorough testing on a variety of real-world catastrophe datasets to demonstrate the applicability of our planned image processing pipeline. We've also performed a series of stress tests to figure out how throughput and latency affect particular components. Our system's first performance findings demonstrate that it can assist official crisis responders in processing photos from social media in real-time.

We think that the suggested real-time online image processing pipeline will aid in the timely and effective extraction of important information from social media picture content. We think that, amongst many other things, the provided image processing pipeline may assist humanitarian organizations in making early decisions by acquiring situational awareness during an ongoing incident or analyzing the level of catastrophe damage.

REFERENCES

1. Afzal, S., & Robinson, P., (2011). Natural affect data: Collection and annotation. In: *New perspectives on Affect and Learning Technologies* (Vol. 1, pp. 55–70). Springer, New York, NY.
2. Alam, F., Imran, M., & Ofli, F., (2017). Image4act: Online social media image processing for disaster response. In: *International conference on advances in social networks analysis and mining (ASONAM)* (Vol. 1, pp. 1–4). Sydney, Australia: IEEE/ACM.
3. Alam, F., Ofli, F., & Imran, M., (2018). Processing social media images by combining human and machine computing during crises. *International Journal of Human–Computer Interaction, 34*(4), 311–327.
4. Al-Ayyoub, M., Jararweh, Y., Benkhelifa, E., Vouk, M., & Rindos, A., (2015). SDsecurity: A software defined security experimental framework. In: *2015 IEEE international conference on communication workshop (ICCW)* (Vol. 1, pp. 1871–1876). IEEE.
5. Amasaki, S., (2019). Exploring preference of chronological and relevancy filtering in effort estimation. In: *International Conference on Product-Focused Software Process Improvement* (Vol. 1, pp. 247–262). Springer, Cham.
6. An, S., Huang, Z., Chen, Y., & Weng, D., (2017). Near duplicate product image detection based on binary hashing. In: *Proceedings of the 2017 International Conference on Deep Learning Technologies* (Vol. 1, pp. 75–80). are features in deep neural networks? In: *Advances in Neural Information Processing Systems* (pp. 3320–3328). Montr'eal, Canada: NIPS.
7. Aroyo, L., & Welty, C., (2015). Truth is a lie: Crowd truth and the seven myths of human annotation (Vol. 36, No. 1, pp. 15–24). AI Magazine.
8. Aswal, N., Sen, S., & Mevel, L., (2022). Switching Kalman filter for damage estimation in the presence of sensor faults. *Mechanical Systems and Signal Processing, 175*(1), 109116.
9. Attari, N., Ofli, F., Awad, M., Lucas, J., & Chawla, S., (2017). Nazr-CNN: Fine-grained classification of UAV imagery for damage assessment. In: *IEEE International Conference on Data Science and Advanced Analytics (DSAA)* (Vol. 1, pp. 1–10). Tokyo, Japan: IEEE.
10. Auffarth, B., López, M., & Cerquides, J., (2010). Comparison of redundancy and relevance measures for feature selection in tissue

classification of CT images. In: *Industrial Conference on Data Mining* (Vol. 1, pp. 248–262). Springer, Berlin, Heidelberg.

11. Babenko, A., Slesarev, A., Chigorin, A., & Lempitsky, V., (2014). Neural codes for image retrieval. In: *European Conference on Computer Vision* (Vol. 1, pp. 584–599). Zurich, Switzerland: Springer.
12. Basbug, S., (2020). Design of circular array with Yagi-Uda corner reflector antenna elements and camera trap image collector application. *Progress In Electromagnetics Research M, 94*(1), 51–59.
13. Bin, Y., Zhou, K., Lu, H., Zhou, Y., & Xu, B., (2017). Training data selection for cross-project defection prediction: Which approach is better?. In: *2017 ACM/IEEE International Symposium on Empirical Software Engineering and Measurement* (ESEM) (Vol. 1, pp. 354–363). IEEE.
14. Bozdag, E., (2013). Bias in algorithmic filtering and personalization. *Ethics and Information Technology, 15*(3), 209–227.
15. Burger, S., Weilhammer, K., Schiel, F., & Tillmann, H. G., (2000). Verbmobil data collection and annotation. In: *Verbmobil: Foundations of Speech-to-Speech Translation* (Vol. 1, pp. 537–549). Springer, Berlin, Heidelberg.
16. Chen, T., Lu, D., Kan, M. Y., & Cui, P., (2013). Understanding and classifying image tweets. In: *ACM International Conference on Multimedia* (Vol. 1, pp. 781–784). Barcelona, Spain: ACM.
17. Chen, Z., & Freire, J., (2020). Proactive discovery of fake news domains from real-time social media feeds. In: *Companion Proceedings of the Web Conference 2020* (Vol. 1, pp. 584–592).
18. Chum, O., Philbin, J., & Zisserman, A., (2008). Near duplicate image detection: Min-hash and tf-idf weighting. In: *British Machine Vision Conference (BMVC)* (Vol. 1, pp. 1–10). Leeds, UK: The University of Leeds. Colorado, USA: IEEE.
19. Cresci, S., Tesconi, M., Cimino, A., & Dellórletta, F., (2015). A linguistically- driven approach to cross-event damage assessment of natura disasters from social media messages. In: *ACM International Conference on World Wide Web* (Vol. 1, pp. 1195–1200). Florence, Italy: ACM.
20. Csurka, G., Dance, C. R., Fan, L., Willamowski, J., & Bray, C., (2004). Visual categorization with bags of key- points. In: *Workshop on Statistical Learning in Computer vision, ECCV* (Vol. 1, pp. 59–74). Prague, Czech Republic: Springer.

21. Daly, S., & Thom, J. A., (2016). Mining and classifying image posts on social media to analyze fires. In: *International Conference on Information Systems for Crisis Response and Management (ISCRAM)* (Vol. 1, pp. 1–14). Rio de Janeiro, Brazil: Scopus.
22. Delpueyo, D., Balandraud, X., & Grédiac, M., (2013). Heat source reconstruction from noisy temperature fields using an optimized derivative gaussian filter. *Infrared Physics & Technology, 60*, 312–322.
23. Ding, Y., Xiao, J., & Yu, J., (2011). Importance filtering for image retargeting. In: *CVPR 2011* (Vol. 1, pp. 89–96). IEEE.
24. Donahue, J., Jia, Y., Vinyals, O., Hoffman, J., Zhang, N., Tzeng, E., & Darrell, T., (2014). Decaf: A deep convolutional activation feature for generic visual recognition. In: *International Conference on Machine Learning* (Vol. 1, pp. 647–655). Beijing, China: International Machine Learning Society (IMLS).
25. Dong, W., Wang, Z., Charikar, M., & Li, K., (2012). High- confidence near-duplicate image detection. In: *Proceedings of the 2nd ACM International Conference on Multimedia Retrieval* (Vol. 1, p. 1). Hong Kong: ACM.
26. Elmgerbi, A., Thonhauser, G., Fine, A., Hincapie, R. E., & Borovina, A., (2021). Experimental approach for assessing filter-cake removability derived from reservoir drill-in fluids. *Journal of Petroleum Exploration and Production Technology, 11*(11), 4029–4045.
27. Everingham, M., Van, G. L., Williams, C. K. I., Winn, J., & Zisserman, A., (2010). The pascal visual object classes (VOC) challenge. *International Journal of Computer Vision, 88*(2), 303–338.
28. Feng, T., Hong, Z., Fu, Q., Ma, S., Jie, X., Wu, H., & Tong, X., (2014). Application and prospect of a high- resolution remote sensing and geo-information system in estimating earthquake casualties. *Natural Hazards and Earth System Sciences, 14*(8), 2165–2178.
29. Fernandez, G. J., Kerle, N., & Gerke, M., (2015). UAV-based urban structural damage assessment using object-based image analysis and semantic reasoning. *Natural Hazards and Earth System Science, 15*(6), 1087–1101.
30. Frith, C., Perry, R., & Lumer, E., (1999). The neural correlates of conscious experience: An experimental framework. *Trends in Cognitive Sciences, 3*(3), 105–114.

31. Gay, G., Menzies, T., Cukic, B., & Turhan, B., (2009). How to build repeatable experiments. In: *Proceedings of the 5th International Conference on Predictor Models in Software Engineering* (Vol. 1, pp. 1–9).

32. Girshick, R. B., Donahue, J., Darrell, T., & Malik, J., (2014). Rich feature hierarchies for accurate object detection and semantic segmentation. In: *IEEE Conference on Computer Vision and Pattern Recognition* (Vol. 1, pp. 580–587). Columbus, Ohio: IEEE.

33. Gong, Y., Pawlowski, M., Yang, F., Brandy, L., Bourdev, L., & Fergus, R., (2015). Web scale photo hash clustering on a single machine. In: *Proceedings of the IEEE Conference on Computer Vision and Pattern Recognition* (Vol. 1, pp. 19–27). Hawaii, USA: IEEE. group. Darmstadt, Germany: IEEE.

34. Hamming, R. W., (1950). Error detecting and error correcting codes. *The Bell System Technical Journal, 29*(2), 147–160.

35. He, K., Zhang, X., Ren, S., & Sun, J., (2016). Deep residual learning for image recognition. In: *IEEE Conference on Computer Vision and Pattern Recognition (CVPR)* (Vol. 1, pp. 4–9). Washington, DC: IEEE.

36. Hughes, A. L., & Palen, L., (2009). Twitter adoption and use in mass convergence and emergency events. *International Journal of Emergency Management, 6*(3, 4), 248–260.

37. Imran, M., Castillo, C., Diaz, F., & Vieweg, S., (2015). Processing social media messages in mass emergency: A survey. *ACM Computing Surveys, 47*(4), 67–69.

38. Imran, M., Castillo, C., Lucas, J., Meier, P., & Vieweg, S., (2014). AIDR: Artificial intelligence for disaster response. In: *ACM International Conference on World Wide Web* (Vol. 1, pp. 159–162). Seoul, Republic of Korea: ACM.

39. Imran, M., Elbassuoni, S. M., Castillo, C., Diaz, F., & Meier, P., (2013). Extracting information nuggets from disaster-related messages in social media. In: International conference on information systems for crisis response and management. Baden-Baden, Germany: Scopus. In: *IEEE Conference on Computer Vision and Pattern Recognition* (Vol. 1, pp. 1717–1724). Columbus, Ohio: IEEE.

40. Itaki, T., Taira, Y., Kuwamori, N., Saito, H., Hoshino, T., & Ikehara, M., (2018). A new tool for microfossil analysis in the southern ocean-automatic image collector equipped with deep learning. In: *AGU Fall Meeting Abstracts* (Vol. 2018, pp. PP23E-1550).

41. Jégou, H., Douze, M., Schmid, C., & Pérez, P., (2010). Aggregating local descriptors into a compact image representation. *IEEE Conference on Computer Vision and Pattern Recognition (CVPR)* (Vol. 1, pp. 3304–3311). San Francisco, USA: IEEE.
42. Jiang, T., & Liu, X. J., (2012). Remote backup system based on data de-duplication. *Computer Engineering and Design, 33*(12), 4546–4550.
43. Kawata, K., Amasaki, S., & Yokogawa, T., (2015). Improving relevancy filter methods for cross-project defect prediction. In: *2015 3rd International Conference on Applied Computing and Information Technology/2nd International Conference on Computational Science and Intelligence* (Vol. 1, pp. 2–7). IEEE.
44. Ke, Y., Sukthankar, R., & Huston, L., (2004). An efficient parts-based near-duplicate and sub-image retrieval system. In: *ACM International Conference on Multimedia* (Vol. 1, pp. 869–876). New York, NY: ACM.
45. Koppula, H. S., Leela, K. P., Agarwal, A., Chitrapura, K. P., Garg, S., & Sasturkar, A., (2010). Learning URL patterns for webpage de-duplication. In: *Proceedings of the Third ACM International Conference on Web Search and Data Mining* (Vol. 1, pp. 381–390).
46. Krallinger, M., Rabal, O., Leitner, F., Vazquez, M., Salgado, D., Lu, Z., & Valencia, A., (2015). The CHEMDNER corpus of chemicals and drugs and its annotation principles. *Journal of Cheminformatics, 7*(1), 1–17.
47. Krizhevsky, A., Sutskever, I., & Hinton, G. E., (2012). ImageNet classification with deep convolutional neural networks. In: *Advances in Neural Information Processing Systems* (Vol. 1, pp. 1097–1105). Lake Tahoe, USA: ACM.
48. Kudela, P., Radzieński, M., & Ostachowicz, W., (2015). Identification of cracks in thin-walled structures by means of wavenumber filtering. *Mechanical Systems and Signal Processing, 50*(1), 456–466.
49. Kudela, P., Radzienski, M., & Ostachowicz, W., (2018). Impact induced damage assessment by means of lamb wave image processing. *Mechanical Systems and Signal Processing, 102*(1), 23–36.
50. Kuminski, E., George, J., Wallin, J., & Shamir, L., (2014). Combining human and machine learning for morphological analysis of galaxy images (Vol. 126, No. 944, p. 959). Publications of the Astronomical Society of the Pacific.

51. Labani, M., Moradi, P., Ahmadizar, F., & Jalili, M., (2018). A novel multivariate filter method for feature selection in text classification problems. *Engineering Applications of Artificial Intelligence, 70*, 25–37.
52. Lagerstrom, R., Arzhaeva, Y., Szul, P., Obst, O., Power, R., Robinson, B., & Bednarz, T., (2016). Image classification to support emergency situation awareness. *Frontiers in Robotics and AI, 3*(1), 54–60.
53. Lavrac, N., Ganberger, D., & Turney, P., (1998). A relevancy filter for constructive induction. *IEEE Intelligent Systems and their Applications, 13*(2), 50–56.
54. Lazebnik, S., Schmid, C., & Ponce, J., (2006). Beyond bags of features: Spatial pyramid matching for recognizing natural scene categories. In: *IEEE Conference on Computer Vision and Pattern Recognition* (Vol. 2, p. 2169–2178). New York, NY: IEEE.
55. Yosinski, J., Clune, J., Bengio, Y., & Lipson, H. (2014). How transferable are features in deep neural networks?. Advances in neural information processing systems, 27..
56. Lee, D. C., Ke, Q., & Isard, M., (2010). Partition min-hash for partial duplicate image discovery. In: *European Conference on Computer Vision (ECCV)* (pp. 648–662). Crete, Greece: Springer.
57. Lee, J. S., (1981). Refined filtering of image noise using local statistics. *Computer Graphics and Image Processing, 15*(4), 380–389.
58. Lei, Y., Qiu, G., Zheng, L., & Huang, J., (2014). Fast near-duplicate image detection using uniform randomized trees. *ACM Trans. Multimedia Comput. Commun. Appl., 10*(4), 1–15.
59. Lei, Y., Wang, Y., & Huang, J., (2011). Robust image hash in radon transform domain for authentication. *Signal Processing: Image Communication, 26*(6), 280–288.
60. Lu, G., Debnath, B., & Du, D. H., (2011). A forest-structured bloom filter with flash memory. In: *2011 IEEE 27th Symposium on Mass Storage Systems and Technologies (MSST)* (Vol. 1, pp. 1–6). IEEE.
61. Lu, G., Jin, Y., & Du, D. H., (2010). Frequency based chunking for data de-duplication. In: *2010 IEEE International Symposium on Modeling, Analysis and Simulation of Computer and Telecommunication Systems* (Vol. 1, pp. 287–296). IEEE.
62. Ma, L., Zhen, C., Zhao, B., Ma, J., Wang, G., & Liu, X., (2010). Towards fast de-duplication using low energy coprocessor. In: *2010*

IEEE Fifth International Conference on Networking, Architecture, and Storage (Vol. 1, pp. 395–402). IEEE.

63. Maynard, D., Roberts, I., Greenwood, M. A., Rout, D., & Bontcheva, K., (2017). A framework for real-time semantic social media analysis. *Journal of Web Semantics, 44*(1), 75–88.
64. McNitt-Gray, M. F., Armato, III. S. G., Meyer, C. R., Reeves, A. P., McLennan, G., Pais, R. C., & Clarke, L. P., (2007). The lung image database consortium (LIDC) data collection process for nodule detection and annotation. *Academic Radiology, 14*(12), 1464–1474.
65. Melenli, S., & Topkaya, A., (2020). Real-time maintaining of social distance in COVID-19 environment using image processing and big data. In: *The International Conference on Artificial Intelligence and Applied Mathematics in Engineering* (Vol. 1, pp. 578–589). Springer, Cham.
66. Meng, J., & Paulson, L. C., (2009). Lightweight relevance filtering for machine-generated resolution problems. *Journal of Applied Logic, 7*(1), 41–57.
67. Nelson, A., Menzies, T., & Gay, G., (2011). Sharing experiments using open□source software. *Software: Practice and Experience, 41*(3), 283–305.
68. Nguyen, D. T., Alam, F., Ofli, F., & Imran, M., (2017). Automatic image filtering on social networks using deep learning and perceptual hashing during crises. In: *International Conference on Information Systems for Crisis Response and Management (ISCRAM)* (Vol. 1, pp. 3–9). Albi, France: Scopus.
69. Nguyen, D. T., Ofli, F., Imran, M., & Mitra, P., (2017). Damage assessment from social media imagery data during disasters. In: *International Conference on Advances in Social Networks Analysis and Mining (ASONAM)* (Vol. 1, pp. 1–8). Sydney, Australia: IEEE/ACM.
70. Ofli, F., Meier, P., Imran, M., Castillo, C., Tuia, D., & Rey, N. O., (2016). Combining human computing and machine learning to make sense of big (aerial) data for disaster response. *Big Data, 4*(1), 47–59.
71. Onno, P., & Guillemot, C. M., (1993). Tradeoffs in the design of wavelet filters for image compression. In: *Visual Communications and Image Processing 93* (Vol. 2094, pp. 1536–1547). International Society for Optics and Photonics.

72. Oquab, M., Bottou, L., Laptev, I., & Sivic, J., (2014). *Learning and Transferring Mid-Level Image Representations Using Convolutional Neural Networks* (Vol. 1, pp. 5–9).
73. Ostachowicz, W., Kudela, P., & Radzienski, M., (2013). Guided wavefield images filtering for damage localization. In: *Key Engineering Materials* (Vol. 558, pp. 92–98). Trans Tech Publications Ltd.
74. Ozbulak, G., Aytar, Y., & Ekenel, H. K., (2016). How transferable are CNN-based features for age and gender classification? In: *International Conference of the Biometrics Special Interest* (Vol. 1, pp. 2–9).
75. Patil, M., & Dongre, N., (2020). Melanoma detection using HSV with SVM classifier and de-duplication technique to increase efficiency. In: *International Conference on Computing Science, Communication and Security* (Vol. 1, pp. 109–120). Springer, Singapore.
76. Pesaresi, M., Gerhardinger, A., & Haag, F., (2007). Rapid damage assessment of built-up structures using VHR satellite data in tsunami-affected areas. *International Journal of Remote Sensing, 28*(13, 14), 3013–3036.
77. Peters, F., Menzies, T., & Marcus, A., (2013). Better cross company defect prediction. In: *2013 10th Working Conference on Mining Software Repositories (MSR)* (Vol. 1, pp. 409–418). IEEE.
78. Peters, R., & Joao, P. D. A., (2015). Investigating images as indicators for relevant social media messages in disaster management. In: *International Conference on Information Systems for Crisis Response and Management* (Vol. 1, pp. 1–9). Kristiansand, Norway: Scopus. recognition. Boston, USA: IEEE.
79. Petrillo, M., & Baycroft, J., (2010). *Introduction to Manual Annotation* (Vol. 1, pp. 1–7). Fairview research.
80. Quirin, A., Abascal-Mena, R., & Sèdes, F., (2018). Analyzing polemics evolution from twitter streams using author-based social networks. *Computación y Sistemas, 22*(1), 35–45.
81. Rahman, M. M., Nathan, V., Nemati, E., Vatanparvar, K., Ahmed, M., & Kuang, J., (2019). Towards reliable data collection and annotation to extract pulmonary digital biomarkers using mobile sensors. In: *Proceedings of the 13th EAI International Conference on Pervasive Computing Technologies for Healthcare* (Vol. 1, pp. 179–188).
82. Rashtchian, C., Young, P., Hodosh, M., & Hockenmaier, J., (2010). Collecting image annotations using amazon's mechanical Turk. In:

Proceedings of the NAACL HLT 2010 Workshop on Creating Speech and Language Data with Amazon's Mechanical Turk (Vol. 1, pp. 139–147).

83. Reuter, C., Ludwig, T., Kaufhold, M. A., & Pipek, V., (2015). Xhelp: Design of a cross-platform social-media application to support volunteer moderators in disasters. In: *ACM Conference on Human Factors in Computing Systems* (Vol. 1, pp. 4093–4102). Seoul, Republic of Korea: ACM.
84. Rivest, R., (1992). *The MD5 Message-Digest Algorithm*. RFC 1321, MIT Laboratory for Computer Science and RSA Data Security.
85. Rudra, K., Banerjee, S., Ganguly, N., Goyal, P., Imran, M., & Mitra, P., (2016). Summarizing situational tweets in crisis scenario. In: *ACM Conference on Hypertext and Social Media* (Vol. 1, pp. 137–147). Nova Scotia, Canada: ACM.
86. Rui, Y., Huang, T. S., & Mehrotra, S., (1997). Content- based image retrieval with relevance feedback in mars. *International Conference on Image Processing* (Vol. 2, pp. 815–818). California, USA: IEEE.
87. Russakovsky, O., Deng, J., Su, H., Krause, J., Satheesh, S., Ma, S., & Fei-Fei, L., (2015). ImageNet large scale visual recognition challenge. *International Journal of Computer Vision, 115*(3), 211–252.
88. Sermanet, P., Eigen, D., Zhang, X., Mathieu, M., Fergus, R., & LeCun, Y., (2013). *Overfeat: Integrated Recognition, Localization and Detection Using Convolutional Networks, 1*, 2–9.
89. Shi, S., Wang, Q., Xu, P., & Chu, X., (2016). *Benchmarking State-of-the Art Deep Learning Software Tools* (3rd edn., pp. 4–9).
90. Siddiquie, B., Feris, R. S., & Davis, L. S., (2011). Image ranking and retrieval based on multi-attribute queries. *IEEE Conference on Computer Vision and Pattern Recognition (CVPR)* (Vol. 1, pp. 801–808).
91. Simonyan, K., & Zisserman, A., (2014). *Very Deep Convolutional Networks for Large-Scale Image Recognition* (2rd edn., pp. 1–6).
92. Singh, A., Shekhar, S., & Jalal, A. S., (2012). Semantic based image retrieval using multi-agent model by searching and filtering replicated web images. In: *2012 World Congress on Information and Communication Technologies* (Vol. 1, pp. 817–821). IEEE.

93. Speck, J. A., Schmidt, E. M., Morton, B. G., & Kim, Y. E., (2011). A comparative study of collaborative vs. traditional musical mood annotation. In: *ISMIR* (Vol. 104, pp. 549–554).
94. Starbird, K., Palen, L., Hughes, A. L., & Vieweg, S., (2010). Chatter on the red: What hazards threat reveals about the social life of microblogged information. In: *ACM Conference on Computer Supported Cooperative Work* (Vol. 1, pp. 241–250). Savannah, USA: ACM.
95. Sulaiman, M. S., Nordin, S., & Jamil, N., (2017). An object properties filter for multi-modality ontology semantic image retrieval. *Journal of Information and Communication Technology, 16*(1), 1–19.
96. Szegedy, C., Liu, W., Jia, Y., Sermanet, P., Reed, S., Anguelov, D., & Rabinovich, A., (2015). Going deeper with convolutions. In: *Proceedings of the IEEE Conference on Computer Vision and Pattern* (Vol. 1, pp. 3–8).
97. Szegedy, C., Vanhoucke, V., Ioffe, S., Shlens, J., & Wojna, Z., (2015). *Rethinking the Inception Architecture for Computer Vision* (4th edn., pp. 2–9).
98. Tseng, C. M., Ciou, J. R., & Liu, T. J., (2014). A cluster-based data de-duplication technology. In: *2014 Second International Symposium on Computing and Networking* (Vol. 1, pp. 226–230). IEEE.
99. Turker, M., & San, B. T., (2004). Detection of col- lapsed buildings caused by the 1999 Izmit, turkey earthquake through digital analysis of post event aerial photographs. *International Journal of Remote Sensing, 25*(21), 4701–4714.
100. Tzeng, Y. C., Chiu, S. H., Chen, D., & Chen, K. S., (2007). Change detections from SAR images for damage estimation based on a spatial chaotic model. In: *2007 IEEE International Geoscience and Remote Sensing Symposium* (Vol. 1, pp. 1926–1930). IEEE.
101. Wu, Z., Ke, Q., Isard, M., & Sun, J., (2009). Bundling features for large scale partial-duplicate web image search. In: *Computer Vision and Pattern Recognition, 2009-CVPR 2009: IEEE conference* (Vol. 1, pp. 25–32). Miami, USA: IEEE.
102. Xiai, Y., Dafang, Z., & Boyun, Z., (2016). Study on data de-duplication based on bloom filter in fault tolerant erasure codes. *Journal of Computational and Theoretical Nanoscience, 13*(1), 179–185.

103. Xu, K., Ba, J., Kiros, R., Cho, K., Courville, A., Salakhudi-Nov, R., & Bengio, Y., (2015). Show, attend and tell: Neural image caption generation with visual attention. In: *International Conference on Machine Learning* (Vol. 1, pp. 2048–2057). Lille, France: International Machine.

104. Yanai, K., (2007). Image collector III: A web image-gathering system with bag-of-key points. In: *Proceedings of the 16th international conference on World Wide Web* (Vol. 1, pp. 1295, 1296).

105. Yang, G. Z., Burger, P., Firmin, D. N., & Underwood, S. R., (1996). Structure adaptive anisotropic image filtering. *Image and Vision Computing, 14*(2), 135–145.

106. Yuen, J., Russell, B., Liu, C., & Torralba, A., (2009). LabelMe video: Building a video database with human annotations. In: *2009 IEEE 12th International Conference on Computer Vision* (Vol. 1, pp. 1451–1458). IEEE.

107. Zagoruyko, S., & Komodakis, N., (2016). *Wide Residual Networks* (2nd edn., pp. 2–8).

108. Zauner, C., (2010). *Implementation and Benchmarking of Perceptual Image Hash Functions*. In: PhD dissertation.

109. Zeiler, M. D., & Fergus, R., (2014). Visualizing and understanding convolutional networks. In: David, F., Tomas, P., Bernt, S., & Tinne, T., (eds.), *European Conference on Computer Vision (ECCV14)* (Vol. 1, pp. 818–833). Zurich, Switzerland.

110. Zhang, M., & Gunturk, B. K., (2008). Multiresolution bilateral filtering for image denoising. *IEEE Transactions on Image Processing, 17*(12), 2324–2333.

111. Zheng, S., Song, Y., Leung, T., & Goodfellow, I., (2016). Improving the robustness of deep neural networks via stability training. In: *Proceedings of the IEEE Conference on Computer Vision and Pattern Recognition* (pp. 4480–4488). Washington, DC: IEEE.

CHAPTER 3

ONLINE SOCIAL MEDIA IMAGE PROCESSING FOR DISASTER MANAGEMENT

CONTENTS

3.1. INTRODUCTION

In today's disaster response, postings on social media are becoming more common. The circumstances and indications included in photographs uploaded on social media, in addition to textual material, play a crucial role in selecting proper rescue operations to a specific disaster. In this chapter, a disaster category for disaster response is suggested and the same taxonomy is used to automate the disaster response decision-making process using an emergency service workflow and deep-learning-based picture identification and object classification algorithms (Layek et al., 2019; Ofli et al., 2020). To ensure that the disaster classification was thorough and accurate, the card sorting technique was used. To evaluate catastrophe-related photos and determine disaster kinds and important cues, VGG-16 and You Only Look Once (YOLO) algorithms were used. Additionally, the intermediate outputs were aligned with the help of decision tables and analytic hierarchy processes (AHP) was applied to map the disaster-associated picture into the disaster category and select an acceptable kind of rescue operation for a specific disaster. Earthquake, Typhoon, and Hurricane were used to evaluate the proposed technique. Using YOLOv4, 96% of photos were accurately classified according to disaster taxonomy. An incremental training strategy can help enhance accuracy even more. The approach proposed has the potential to be beneficial in real disaster management because of the usage of cloud-based learning algorithms in processing of image. Other spatiotemporal characteristics retrieved from multimedia content shared on social media can be used to improve the algorithms and the suggested emergency management pipeline (Imran et al., 2018; 2020).

The increased usage of social media sites like Twitter, Instagram, and Facebook during mass cases of emergency caused by natural or man-made catastrophes has offered a number of chances for users to acquire accessibility to timely and relevant insights. During such occurrences, onlookers, and impacted individuals upload current updates, like reports of injured or deceased individuals, damage to infrastructure, pleas for immediate necessities like food, water, and shelter, and contribution offers, among other things. This social network data comes in a number of formats, including text messages, photos, and videos. Quick access to these turn of events and time-sensitive updates via social media sites is valuable for a wide range of real-world applications and can assist fulfill a variety of information requirements (Imran et al., 2017; Nunavath and Goodwin, 2018).

This work focuses on quick emergency response and management, among many other applications. Formal humanitarian groups, law enforcement authorities, as well as other voluntary associations all seek quick information in order to assess the situation and prepare relief activities. Social media data is critical for an effective emergency response, according to research (Imran et al., 2018). Furthermore, to analyze this data, a number of approaches centered on machine learning and artificial intelligence (AI) have been created. Most of these researches, on the other hand, are primarily concerned with assessing textual information, neglecting the plethora of data supplied by visual content. A solution is described in this chapter that addresses this constraint by adding an actual social media picture processing pipeline to aid humanitarian groups in emergency response and management activities (Li et al., 2017).

Real-time analysis of high-velocity and bulk social media visual data is a tough endeavor. In addition, a substantial proportion of social media photographs contain irrelevant and redundant data, resulting in a low signal-to-noise ratio (SNR). Figure 3.1 displays instances of related pictures in the 1st 2 columns and unrelated images (such as cartoons, and celebrities) in the 3rd.

Figure 3.1. Related (1st 2 columns) and inappropriate images (3rd column) gathered during different catastrophe on twitter.

Source: https://www.researchgate.net/figure/Relevant-first-two-columns-and-irrelevant-images-third-column-collected-during_Fig1_321064926.

Before performing an in-depth study of the graphical data of social media to retrieve relevant information for relief groups, the basic social media imagery information must be done to remove this background noise. In this research, an online image analysis pipeline including de-duplication and relevance processing modules to gather and screen social media photos in real-time throughout a disaster is presented.

To evaluate the high-velocity stream of Twitter, the system uses a combination of machine learning and man-made computations approaches. During emergencies, Stand-By-Task-Force (SBTF) volunteers are engaged to assist with image tagging, which would be used to teach machine learning models for particular humanitarian use situations, such as damage evaluation. Before using humans for picture annotation, it's critical to remove similar and unnecessary images to save money and time on the crowdsourcing budget. Such that, human volunteers shouldn't spend their time continually tagging photographs that are useless or duplicate. Deep neural networks and perceptual hashing algorithms are used to detect similar and unnecessary images. The decoded stream of photos is also evaluated to determine the extent of damage to infrastructure visible in the photographs (Asif et al., 2021; Khatoon et al., 2022).

Occasionally, calamities occur with no or little warning, resulting in difficulties riddled with uncertainty and demanding us to make snap decisions depending on little knowledge. In this age of pervasive technology and social media, we are often inundated with data that we can't quickly comprehend or filter. The commencement of a disaster necessitates a prompt response to ensure the safety of impacted individuals and safeguard susceptible locations and other assets. In such a case, the instantaneous and continuous flow of information from the impacted areas assists humanitarian groups in providing aid to affected individuals and specific areas. Various researches have demonstrated that social media information is valuable for disaster management and rescue operations. The Emergency Decision Making (EDM) has an important role in natural calamities to aid the relief causes of the relevant organizations during disaster response operations. However, the EDM demands prompt gathering of pertinent data. The majority of prior research indicates that textual data from people on social media aids emergency management during all stages of disaster management: the quiet before the crisis, even during disaster, the peak, the decrease, and the return to normalcy (Nguyen et al., 2017; Hayes and Kelly, 2018). Almost every social networking site allows users to submit photos and videos from their accounts. The majority of social media members find it simpler to snap and share images of the issue than to compose lengthy texts. Twitter and other social media networks offer APIs to access, collect, and download user-uploaded photographs and data. Consequently, we are advised to utilize social media imagery to recognize calamities and disaster-associated cues in posts on social media. The successful collection and integration of information hidden in social media photographs will benefit in many facets of disaster management, such as rapid reaction, rescue, and help operations, and the generation of timely alerts or advisories that might save people and assets (Palen and Hughes, 2018).

Although earlier studies have demonstrated the effectiveness of social media in crisis response and management, it is worth noting that posts on social media aren't always of great quality, precise, or timely. Dwarakanath et al. (2021), for example, acknowledged the importance of social media in disaster-association news communication and information dissemination, as well as appropriate methodologies for content filtering, activity recognition, and summarization. Dong et al. (Dong et al., 2021) looked at posts on Twitter and Flickr that included photographs of the impacted areas. They found that almost all posts on social media during the incident have visual signals relating to supplies and food. (Ofli et al., 2016) looked into image-related geographical data to see if the photographs provided were associated to a tragedy. (Kim and Hastak, 2018) evaluated the spatial information of social networking photographs connected to fire threats in a similar way. Using machine learning approaches, Nguyen et al. (Nguyen et al., 2017) studied social media photographs to evaluate the level of damage following natural calamities. Previous research has shown that individuals' social media photographs are a valuable and comprehensive source of data / information.

Numerous studies have added to disaster picture collections like floods, and earthquakes, which are useful for disaster categorization. Nevertheless, the visual data encoded in social media photographs can be made more valuable by identifying the genuine emergency indicators that humans detect when viewing a picture. To achieve this, additional labeling of these image datasets is required. In media platforms crowdsourcing, the information is typically available in vast quantities; hence, it is difficult for an individual or a team to filter loads of social networking photographs manually and quickly identify meaningful data (i.e., significant indicators) within them. Thus, automating the identification of relevant disaster-associated objects and their relevant data in social media photographs, like destruction, death, causality, people affected, livestock, or supplies, etc., aids rescuers, police, and rescue managers in the EDM. Nevertheless, neither of the prior research investigates the following in detail: (a) identifying the emergency signals from the pictures, (b) automating the decision making for tracking down the proper emergency types and supervisory bodies for given image-driven signals, and (c) categorizing the emergency signals and tragedy objects in the data sets. This chapter focuses on the aforementioned characteristics of social network image processing (Ofli et al., 2016; Dominguez et al., 2021).

The goal of this study is to tackle the problem of sorting through a large range of social media photographs depends on the nature of crisis and the appropriate emergency reaction. A new method is presented for automatically recognizing emergency message from photographs shared on social media in the aftermath of a tragedy. The disaster classification of rescue operations is created to help automate media platforms picture analysis in identifying proper emergency responses. Many previous research looked at catastrophe taxonomies from the operational and theoretical perspectives of disasters. However, because lower-level data on the crisis response function are required to construct the automation pipeline, the current taxonomies are more specific to our requirements. As a result, it is necessary to investigate which visual clues can help to speed up the emergency reaction. The disaster item and feature tagged picture dataset to suit the disaster response taxonomy is also created (Laituri and Kodrich, 2008; Foresti et al., 2015).

The analysis of the system on current disaster datasets and its implementation throughout a real-world major disaster (Cyclone Debbie, which struck Queensland, Australia in March 2017) show its efficacy.

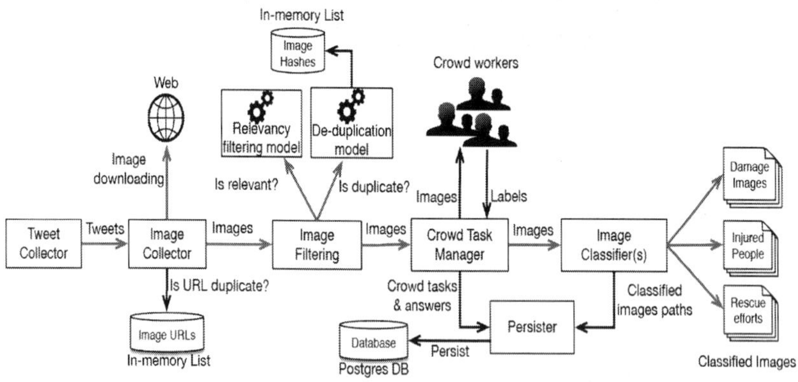

Figure 3.2. Architecture of an online image analysis pipeline for media platforms.

Source; https://www.researchgate.net/figure/Relevant-first-two-columns-and-irrelevant-images-third-column-collected-during_Fig1_321064926.

3.2. IMAGE PROCESSING PIPELINE

The parts of the suggested Image4Act system are visualized in Figure 3.2. Redis Channels2 are used to communicate flow of data between the modules. Furthermore, every module contains a set of RESTFul APIs to permit external interactions and, if necessary, set values of parameter. The red arrows in the diagram indicate live streams of data, while the black arrows indicate non-streaming interactions. The Java Enterprise Edition (J2EE) computer language is used to build the system (Chaudhuri and Bose, 2019; Freitas et al., 2020).

3.2.1. Tweet Collector

This part collects live tweets via Twitter's streaming API3. A user defines keywords, tags, geographic bounding boxes, and Twitter users to build a database for a certain event. In the event of the geographically bounding box selection, solely geo-tagged tweets are gathered; nevertheless, both the keywords and geographic bounding box choices can be used to retrieve tweets that meet any of the two factors (Wang and Ye, 2018).

3.2.2. Image Collector

This part consumes tweets gathered by the Tweet Collector in order to obtain picture URLs from the gathered tweets. For de-duplication objectives, every URL is compared to in key-value pair library that stores distinct URLs. Images with distinct URLs are retrieved from the Web and posted to the Redis channel in a pipeline. This channel's subscribers receive photographs as soon as they are uploaded (Xu et al., 2016).

3.2.3. Image Filtering

The context of relevance differs throughout crisis occurrences, humanitarian groups, including within a long-running crisis, making modeling relevancy a difficult task. Contrariwise, what is judged unnecessary appears consistent throughout calamities and among numerous humanitarian agencies (Imran et al., 2020). In other words, visuals depicting cartoons, celebrities, and commercials are instances of inappropriate content, and are therefore useless for calamity management and response. The Image Filtering part uses deep learning models and perceptual hashing algorithms to identify if a recently arrived image (a) is pertinent for a given crisis response scenario and (b) isn't a duplication of previously gathered photos. Specifically, Convolutional

Neural Networks (CNN), VGG-16 architecture, in conjunction with the DeepLearning4J toolkit are used, to train a relevance model. Perceptual Hashing technique is used to calculate a pHash value for every image, which is subsequently stored in the in-memory library for duplicate recognition. Using the Hamming distance, a freshly received picture pHash is matched to the saved hashes to identify copies and near-duplicates (Li et al., 2018; Ofli et al., 2022).

3.3. CROWD TASK MANAGER

This part is in charge of allocating SBTF participants to picture tagging assignments. An assignment, also known as a classifier, is created by an end-user and includes a set of classifications (e.g., severe damage, slight damage and no damage). A human labeler is shown an image and a set of categories by the Crowd Task Manager. The labeler assigns a suitable label to the picture, which is then used as a training case.

3.3.1. Image Classifiers

The system permits users to specify one or more classifications. A classifier could include two or more classes. The user-defined picture classifiers are trained using human-labeled photos collected from the Crowd Task Manager. Since many investigations in computer vision literature,–for example, have demonstrated that the characteristics learned by CNNs on basic large-scale image recognition tasks (i.e., billions and billions of variables, trained on billions of pictures from the ImageNet dataset [(Ravi et al., 2019) are demonstrated to be interchangeable and efficient when used in certain specific tasks, especially when training set seem to be limited,–as that it's in our case especially during the early stages of collection of data–, transfer learning approach is adopted. The last component of the model to conform to the training set supplied by the user is also adapted, rather than the original categorization of one thousand classes. Therefore, this transfer learning method permits us to transfer the network's characteristics and variables from a broad domain to the narrow one.

3.3.2. Persister

All database-particular actions, like insertion of picture meta-data, storing, and retrieving of predicted results, are handled by this module. Furthermore, it saves machine-tagged photos to the file system.

3.4. DATASET AND SYSTEM ASSESSMENT

Datasets: The suggested Image4Act system was evaluated using images submitted on Twitter throughout four natural calamities. The occurrences and label dissemination acquired using human participants are detailed in Table 3.1. The primary crowdsourcing aim was to identify whether a picture had significant, medium, or no damage. 3,518 photos were selected at random from the extreme and moderate categories and considered appropriate to teach the relevancy filter. The ImageNet 1,000-class VGG-16 model was used to label images in the none class, and the top twelve most often, unrelated categories were chosen with human oversight. The none category yielded 3,518 photos, all of which were considered irrelevant. Asr learning, verification, and test sets, a 60: 20: 20 split were employed (Ravi et al., 2019).

Relevancy Assessment: A transfer learning strategy was adopted to start the VGG-16 network with the parameters of the basic ImageNet model because little labeled data was available. After adjusting the last layer of the network for binary categorization task, the very same network was fine-tuned using our labeled data. AUC of 0.98 was obtained.

Table 3.1. Datasets–NE: Nepal Earthquake, EE: Ecuador Earthquake, TR: Typhoon Ruby, HM: Hurricane Matthew

Class	NE	TR	EE	HM	Total
Severe	8,927	88	955	110	10080
Mild	2257	338	89	94	2778
None	14239	6387	946	132	21704
Total	25423	6813	1990	336	34562

De-duplication Assessment: The effectiveness of de-duplication filter is dependent on the Hamming distance threshold. To get this value, pH ashes of 1,100 randomly selected photos was computed. By manually examining set of images with a distance around 0–20, it was determined that distance less than or equal to 10 is the ideal value of distance at which the system keeps an accuracy of 0.98.

Damage Severity Evaluation: The de-duplication model and relevancy model was used on the original data to produce a cleaned set from which an arbitrary selection of 6k photos (severe=1765, mild=483, none=3751) was chosen for the task of damage evaluation. The damage evaluation model was trained with the help pf 60:20:20 split as learning, verification, and test sets, correspondingly. An average AUC of around 0.72 was obtained, which is within acceptable limits (Reuter et al., 2020).

Deployment during a Real-World Calamity: The method was used in the aftermath of Cyclone Debbie, which hit Queensland, Australia on 28th of March, 2017. Seven thousand tweets with photographs were found among the 76k tweets collected in actual using the tag #CycloneDebbie. To determine whether an arriving image is appropriate, the previously trained relevancy model was engaged. The authors sampled and analyzed five hundred machine-classified photos for their evaluation. For the unrelated and duplicate situations, the accuracy ratings were 0.67 and 0.92, correspondingly. During the current calamity, the system automatically identified valid and invalid photos, as seen in Figure 3.3. It is apparent that the system was able to distinguish between useful and useless images.

There have been few works on crisis response image analysis. The majority of the works in the literature pertain to remote sensing. Liu et al. (Liu et al., 2015) show the usefulness of photos for rescue operations, wherein they acquired photographs with the help of FORMOSAT-2 satellite throughout tsunami in South Asia in 2004. In addition to feature research on satellite image analysis for crisis response. Comparable studies in the field of remote sensing investigation provide proof of damage evaluation using aerial and satellite images of disaster-affected locations (Pohl et al., 2016).

The significance of social media pictures for calamity response has been currently emphasized in Pohl et al., (2016). For the flood catastrophe in Saxony, the authors reviewed messages and tweets from Instagram and Flicker (2013). They discovered that photos inside on-topic messages are more related to the catastrophic occurrence, and the picture content can convey essential information of event. Meier (Meier, 2013) aimed at identifying photographs acquired from media platforms data, such as Flickr, and analyzing whether the fire happened at a specific time and location (Ghosh et al., 2018; Fan et al., 2020). Their research also looked at the photos' spatio-temporal meta-data and found that geotags can help find the fire-hit area.

Figure 3.3. Classified photos sampled throughout cyclone Debbie with data set from the classifier.

Source: https://www.nature.com/articles/s41467–020–19644–6.

3.5. CONCLUSIONS

The system that can accept and evaluate Twitter visual content in real-time to aid humanitarian groups in defining the level of a situation and making smart decisions is presented. With the support of workers, the system included two critical image filtering parts to screen out the noisy information and to assist crisis management in building more fine-grained classifiers, such as damage evaluation from photographs. To show the system's usefulness, it was tested both online and offline during a real-world calamity.

REFERENCES

1. Alam, F., Ofli, F., & Imran, M., (2018). Processing social media images by combining human and machine computing during crises. *International Journal of Human–Computer Interaction, 34*(4), 311–327.
2. Asif, A., Khatoon, S., Hasan, M. M., Alshamari, M. A., Abdou, S., Elsayed, K. M., & Rashwan, M., (2021). Automatic analysis of social media images to identify disaster type and infer appropriate emergency response. *Journal of Big Data, 8*(1), 1–28.
3. Chaudhuri, N., & Bose, I., (2019). Application of image analytics for disaster response in smart cities. In: *Proceedings of the 52nd Hawaii International Conference on System Sciences*.
4. Chaudhuri, N., & Bose, I., (2020). Exploring the role of deep neural networks for post-disaster decision support. *Decision Support Systems, 130*, 113234.
5. Dominguez-Péry, C., Tassabehji, R., Vuddaraju, L. N. R., & Duffour, V. K., (2021). Improving emergency response operations in maritime accidents using social media with big data analytics: A case study of the MV Wakashio disaster. *International Journal of Operations & Production Management*.
6. Dong, Z. S., Meng, L., Christenson, L., & Fulton, L., (2021). Social media information sharing for natural disaster response. *Natural Hazards, 107*(3), 2077–2104.
7. Dwarakanath, L., Kamsin, A., Rasheed, R. A., Anandhan, A., & Shuib, L., (2021). Automated machine learning approaches for emergency response and coordination via social media in the aftermath of a disaster: A review. *IEEE Access, 9*, 68917–68931.
8. Fan, C., Jiang, Y., Yang, Y., Zhang, C., & Mostafavi, A., (2020). Crowd or hubs: Information diffusion patterns in online social networks in disasters. *International Journal of Disaster Risk Reduction, 46*, 101498.
9. Foresti, G. L., Farinosi, M., & Vernier, M., (2015). Situational awareness in smart environments: Socio-mobile and sensor data fusion for emergency response to disasters. *Journal of Ambient Intelligence and Humanized Computing, 6*(2), 239–257.
10. Freitas, D. P., Borges, M. R., & Carvalho, P. V. R. D., (2020). A conceptual framework for developing solutions that organize social

media information for emergency response teams. *Behavior & Information Technology, 39*(3), 360–378.

11. Ghosh, S., Ghosh, K., Ganguly, D., Chakraborty, T., Jones, G. J., Moens, M. F., & Imran, M., (2018). Exploitation of social media for emergency relief and preparedness: Recent research and trends. *Information Systems Frontiers, 20*(5), 901–907.
12. Han, S., Huang, H., Luo, Z., & Foropon, C., (2019). Harnessing the power of crowdsourcing and internet of things in disaster response. *Annals of Operations Research, 283*(1), 1175–1190.
13. Hayes, P., & Kelly, S., (2018). Distributed morality, privacy, and social media in natural disaster response. *Technology in Society, 54*, 155–167.
14. Imran, M., Alam, F., Ofli, F., Aupetit, M., & Bin, K. H., (2017). *Enabling Rapid Disaster Response Using Artificial Intelligence and Social Media*. Qatar Computing Research Institute.
15. Imran, M., Alam, F., Qazi, U., Peterson, S., & Ofli, F., (2020). *Rapid Damage Assessment Using Social Media Images by Combining Human and Machine Intelligence.* arXiv preprint arXiv:2004.06675.
16. Imran, M., Castillo, C., Diaz, F., & Vieweg, S., (2015). Processing social media messages in mass emergency: A survey. *ACM Computing Surveys (CSUR), 47*(4), 1–38.
17. Imran, M., Castillo, C., Diaz, F., & Vieweg, S., (2018). Processing social media messages in mass emergency: Survey summary. In: *Companion Proceedings of the Web Conference 2018* (pp. 507–511).
18. Imran, M., Ofli, F., Caragea, D., & Torralba, A., (2020). Using ai and social media multimodal content for disaster response and management: Opportunities, challenges, and future directions. *Information Processing & Management, 57*(5), 102261.
19. Khatoon, S., Asif, A., Hasan, M. M., & Alshamari, M., (2022). Social media-based intelligence for disaster response and management in smart cities. In: *Artificial Intelligence, Machine Learning, and Optimization Tools for Smart Cities* (pp. 211–235). Springer, Cham.
20. Kim, J., & Hastak, M., (2018). Social network analysis: Characteristics of online social networks after a disaster. *International Journal of Information Management, 38*(1), 86–96.

21. Laituri, M., & Kodrich, K., (2008). On line disaster response community: People as sensors of high magnitude disasters using internet GIS. *Sensors, 8*(5), 3037–3055.

22. Layek, A. K., Poddar, S., & Mandal, S., (2019). Detection of flood images posted on online social media for disaster response. In: *2019 Second International Conference on Advanced Computational and Communication Paradigms (ICACCP)* (pp. 1–6). IEEE.

23. Li, J., He, Z., Plaza, J., Li, S., Chen, J., Wu, H., & Liu, Y., (2017). Social media: New perspectives to improve remote sensing for emergency response. *Proceedings of the IEEE, 105*(10), 1900–1912.

24. Li, X., Caragea, D., Zhang, H., & Imran, M., (2018). Localizing and quantifying damage in social media images. In: *2018 IEEE/ACM International Conference on Advances in Social Networks Analysis and Mining (ASONAM)* (pp. 194–201). IEEE.

25. Meier, P., (2013). Human computation for disaster response. In: *Handbook of Human Computation* (pp. 95–104). Springer, New York, NY.

26. Nguyen, D. T., Alam, F., Ofli, F., & Imran, M., (2017). *Automatic Image Filtering on Social Networks Using Deep Learning and Perceptual Hashing During Crises*. arXiv preprint arXiv:1704.02602.

27. Nunavath, V., & Goodwin, M., (2018). The role of artificial intelligence in social media big data analytics for disaster management-initial results of a systematic literature review. In: *2018 5th International Conference on Information and Communication Technologies for Disaster Management (ICT-DM)* (pp. 1–4). IEEE.

28. Ofli, F., Imran, M., & Alam, F., (2020). Using artificial intelligence and social media for disaster response and management: An overview. *AI and Robotics in Disaster Studies,* 63–81.

29. Ofli, F., Meier, P., Imran, M., Castillo, C., Tuia, D., Rey, N., & Joost, S., (2016). Combining human computing and machine learning to make sense of big (aerial) data for disaster response. *Big Data, 4*(1), 47–59.

30. Ofli, F., Qazi, U., Imran, M., Roch, J., Pennington, C., Banks, V., & Bossu, R., (2022). *A Real-time System for Detecting Landslide Reports on Social Media using Artificial Intelligence*. arXiv preprint arXiv:2202.07475.

31. Palen, L., & Hughes, A. L., (2018). Social media in disaster communication. *Handbook of Disaster Research,* 497–518.
32. Pohl, D., Bouchachia, A., & Hellwagner, H., (2016). Online indexing and clustering of social media data for emergency management. *Neurocomputing, 172*, 168–179.
33. Ravi Shankar, A., Fernandez-Marquez, J. L., Pernici, B., Scalia, G., Mondardini, M. R., & Di Marzo, S. G., (2019). Crowd4Ems: A crowdsourcing platform for gathering and geolocating social media content in disaster response. *International Archives of the Photogrammetry, Remote Sensing and Spatial Information Sciences, 42*, 331–340.
34. Reuter, C., Stieglitz, S., & Imran, M., (2020). Social media in conflicts and crises. *Behavior & Information Technology, 39*(3), 241–251.
35. Wang, Z., & Ye, X., (2018). Social media analytics for natural disaster management. *International Journal of Geographical Information Science, 32*(1), 49–72.
36. Xu, Z., Zhang, H., Sugumaran, V., Choo, K. K. R., Mei, L., & Zhu, Y., (2016). Participatory sensing-based semantic and spatial analysis of urban emergency events using mobile social media. *EURASIP Journal on Wireless Communications and Networking, 2016*(1), 1–9.

CHAPTER 4

UNDERSTANDING AND CLASSIFYING IMAGE TWEETS

CONTENTS

4.1. INTRODUCTION

Because of advancements through bandwidth and imaging technology, the predominant type of social media is now predominantly multimedia rather than text-based. Portrait Twitter posts, which we describe as consumer micro-blog blogs that consist of an embedded Image, are a standard component of consumer content (Hodosh et al., 2013). Image Twitter posts are blog tweets that consist of an embedded photograph. Although it has been possible to connect photos to posts on microblogs for a lot of years now, the problem of publishing of that kind posts has kept them from becoming the majority of microblog posts. Microblogging portals now automatically incorporate Images into the there own posts, a trend that began with China's Sina Weibo and later spread to Twitter and other 3rd services like Instagram (Starbird et al., 2010; Cheng et al., 2011).

Some preliminary studies explain Image tweets. For example, Yu et al. found that Image tweets made up 56.43% of completely micro-blog posts on Sina Weibo in 2011 (Yu et al., 2011). According to the findings of Zhao et al. photo tweets had been shared more frequently and survived for a longer period than text-only posts (Zhao et al., 2013). Therefore, we can assume that users create Image tweets to attract and maintain the attention of their readers.

Even though they are helpful, these observations only begin to solve so many questions that have been raised regarding of new category of public communication. What kinds of Imageries do people typically four insert? Are there noticeable differences between the Images and those that can be found on photos and Images wanting to share on websites such as Flickr? Does the text of tweets which contain Images vary from those of posts which only contain text (Turker & San, 2004)?

We take compiled a set of Images tweets taken from Sina Weibo to provide answers to all these questions. Our contributions are in Castillo (2016):

- dissecting the body of evidence to characterize the Image and text-based content of such tweets, as well as the relation between the variables;
- accumulating annotations for a selected portion of such Image tweets that are included in the text, and;
- Construct an automatic classifier that can differentiate between non-visual and visual tweets, which are both key subclasses of Image tweets (Ofli et al., 2016).

4.2. WHAT ARE IMAGE TWEETS?

To find the answers to such questions, we compiled a collection of tweets that included both content and Image-based tweets. We collected a dataset consisting of 57,595,852 tweets by randomly sampling postings from the public timelines API of Weibo for seven months in the year 2012. To conduct a more in-depth analysis of the tweets, we basically physically annotated a minor segment of the corpus that was 5K in size (Imran et al., 2015).

Image Attributes and Qualities Image Twitter posts have only been the subject of research in a handful of works up until this point. Ishiguro et al. indicated that social curation indication, such as favoriting and clear and direct listing, performs significantly better than feature representation when it comes to predicting the number of views that an Image tweet will receive (Ishiguro et al., 2012). To determine when new events are about to take place based on Image tweets, Wang et al (9developed a hybrid textual and photo segment model (Wang et al., 2016). The Weibo uploader, which implements specific limits and post-processing upon embedded photos, is an idiosyncratic aspect because all photographs deposit in Weibo are treated by the Weibo uploader. (1) Each post may only include a single Image; (2) the EXIF information of each Image is removed; and (3) all photos, except for animated GIFs, are transformed to the JPEG format. 45.1% of the tweets in our database are Images, and among those Images, even now Images predominate: There is a photo in the Jpeg file in 97.5% of all Image tweets, and there is an animated GIF in 2.5% of all Images tweets (Kerle et al., 2019). There are Images of varied quality (both candid and prepared) and changing themes, screenshots, cartoon characters digitalic wallpapers, or other kinds of decorating Images. Figure 4.1 provides 18 samples that highlight the range of Images that may be found in Image tweets. According to the results of the manual assessment for our annotated data, 69.5% for the photos are natural photographs (with those that have been digitally modified), 13.2% are synthetic, and 17.4% remain multi-photo composites. The combination format gets around Weibo restriction of only allowing one Image per post and may be used for several different narrative reasons, such as comparing two or more items or telling a story via a series of Images (Daly and Thom, 2016).

A study of our remarks, who are all customers of the Weibo, revealed that 85% of them personally used theirs cellphone camera for their Image instrument, while just 13.7% used a digital camera. This finding pertains to the sharing behaviors of our annotators. The fact that the majority of Images on Weibo appear to be of poor quality, in contrast to those on Flickr, lends credence to our theory that consumers of Weibo care extra about the photos content than the Image quality (Imran et al., 2013).

Image-based tweets as opposed to text-based posts. By providing answers to the questions "when," "what," and "why," we make an effort to uncover the distinctions between tweets including Images and those containing simply text. When we examine our tweets and plot them according to the hours in which they were posted (as shown in Figure 4.2a), we find that Image tweets are published more often throughout the day. We believe that there are more interesting things and events happening daytime, but we've not been able to verify this hypothesis (Rudra et al., 2016; Katsaggelos et al., 1991). We learned k = 50 latent topics by applying the Latent Dirichlet Allocation (LDA) algorithm to the large, 1 M subclass of the entire datasets to answer the question "what." The value of k was optimized using a held-out set. We found that the ratio of tweets containing Images to those without Images for some of these 50 topics was significantly different than the estimate of 45.1%. Ads and posts or comments on fashion, transport, but also cooking have been adorned with Images, whereas post about feelings and daily normal routine are generally text-only (Tien Nguyen et al., 2017). The list of sample subjects with manual process assigned labels can be found in Figure 4.2(b). Concerning the question "why," numerous studies on tweets consisting only of text have been carried out. The purpose behind posting could be broken down into two categories: social (regular chatter, discussions) and informational (sharing data, reportage news). We can see that the preference for posting photo tweets as opposed to content tweets is strongly linked with the content by looking at the dispersion of LDA topics (Bruni et al., 2014). For instance, to make an advert more informative, tweets typically include an Image of the product being advertised. On the other hand, tweets about mundane aspects of everyday life, such as twitters whose themes are about working and sleeping, are more likely toward only contain text. That provides an answer to the "why" question from a more general perspective; therefore, the next step is to analyze the relationship between the Image and the messages in independent tweets (Peters & De Albuquerque, 2015).

4.3. IMAGE AND TEXT RELATION

Though users can post an Image without accompanying text, it is rare–99.1% of our Image tweets have corresponding text. We want to know why people post both Image and text and the nature of their correlation (Imran et al., 2014).

Two previous studies have attempted to answer this for the general domain. Marsh and White identified 49 (Kerle et al., 2019) relationships, grouping them into three major categories based on the closeness of the relationship. Martinec and Salway studied text-Image relations from two perspectives: status (in terms of relative importance) and logico-semantic (one expands or repeats the same meaning of another). These classifications, although useful, predate Image tweets and do not account for the textual material accessible on social media. In addition, neither technique has been implemented as an automated classifier (Daly and Thom, 2016).

It's normal to think that the two media should work together: an attached Image should accentuate the visual highlights of the post, while the text provides context: time, place, event, or narrative. That is, since the word and the Image are visually associated, we consider both media to be equivalent (Yang et al., 2019). Visually figure reveals tweets (visuals for short) are those in which at least one term or verb matches to a portion of the Image. We saw this trend in the corpus study, but there's also a shockingly high proportion of non-visual Image tweets, where the text and Image had little or no visual correlation. These are difficult to spot just by glancing at the photographs: in Figure 4.1, the left group of 9 images are all from visual tweets, whereas the right group of 9 images are all from non-visual image tweets (Simonyan and Zisserman, 2014). We discovered that the differentiation is dependent on both text and Image content. Figure 4.3 shows two sample visual (top) and two sample non-visual (bottom) tweets. The reasons for using photos in a non-visual tweet are several. The poster included an outside environment in the bottom row, which has no relation to the content but may lure viewers to examine the post. We've noticed that there's a subset of non-visual image tweets that have a common trait: emotive relevance. In these kind of tweets, the content and the image, like in the 3rd step, will be in the same emotional situation (anger, directed at mosquitoes). In such circumstances, the text is the main medium; the image, similarly to emoticon usage, enhances the text's emotive characteristics (Lei et al., 2011; Kasturi et al., 2008). We do, however, note that there is a difference. It's tough to distinguish among emotive and non-visual tweets. Our annotations revealed that the differentiation would be more of a continuum than a binary. As a result, we only examine the binary difference among visual and non-visual classifications, despite the fact that we find it intriguing (Yosinski et al., 2014). The difference among visual and non-visual information is useful in practice. Embedded Images in visual tweets may be used in a content Image search, never from non-visual tweets. For example, the Image in first row of Figure 4.3 would be a good match for the query "sago cream," while the Image in the third line would not appear in a search for "mosquito bites." The categorization may also aid automatic tagging systems in identifying Image-text pairings that do not meet the relevant assumption (i.e., non-visual tweets). Finally, since Images in visual tweets have semantic significance, social media networks may opt to prioritize photos in visual tweets while loading or allocating display (Zauner, 2010).

Figure 4.1. Example of image tweets.

Source: https://optinmonster.com/make-money-on-twitter/ is a good source.

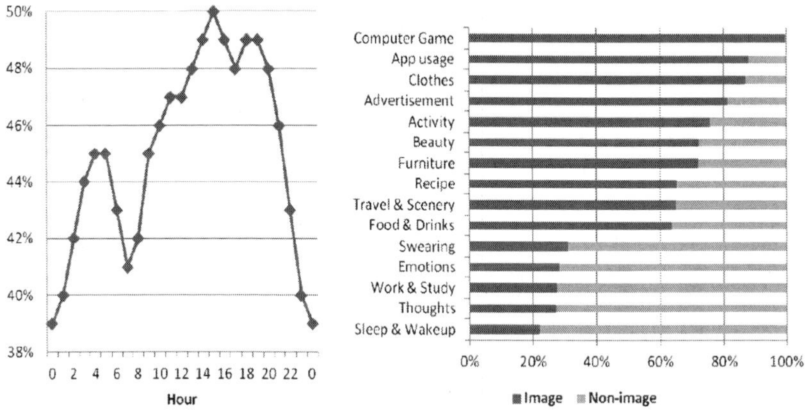

(a) (b)

Figure 4.2. Percentage of image tweets per hour (a). (b) In skewed themes, the ratio of Image to non-image tweets.

Source: https://www.researchgate.net/figure/Distribution-of-tweets-posted-at-different-time-a-The-hourly-pattern-b-The-weakly_Fig3_275974115.

Figure 4.3. Image tweets with their accompanying text, image, and translation. The first two are instances of visual tweets, whereas the latter two are examples of non-visual tweets.

Source: https://www.academia.edu/49030839/Understanding_and_classifying_Image_tweets.

It's difficult to tell the difference between both emotional and nonvisual Twitter posts. Our annotations revealed that the distinction could be more of a continuum than a dual. As a result, we only examine the binary difference among visual and non-visual groups, although we find it intriguing (Ozbulak et al., 2016).

The difference between non-visual and visual information is useful in practice. Embedded Images from visible tweets may be used in a content Image search, and not from non-visual tweets. For example, the Image in the first row of Figure 4.3 would be a good match for such query "sago cream," while the Image in the third row would not appear in a query for "mosquitoes." The categorization might also aid automatic tagging systems in identifying Image-text combinations that do not meet the relevant assumption (i.e., non-visual tweets). Finally, since Images in visual tweets have semantic meaning, social media companies may opt to load or allocate screen real space for visual tweets first (Russakovsky et al., 2015).

4.4. VISUAL/NON-VISUAL CLASSIFICATION

Now we'll use classification tasks to make this difference automatically. We use crowdfunding to create an annotated dataset, and then describe the three kinds of evidence we use for machine learning (Friedman et al., 2005).

Creating a Dataset We used topics from Zhubajie and also graduates from our university to label a random subset of the Image tweets to achieve gold standard annotations. Native Chinese learners and users of microblogs were the subjects of the study. Subjects were asked to classify the photo relationship either as non-visual or visual (Karpathy et al., 2014). Three various topics annotated each photo tweet, with both the simple majority setting the gold standard. In total, we accumulated annotations for 4,811 Image twitter posts annotated by 72 various subjects (also used in our manual analysis in Section 2). There were 3,206 (66.6%) visual Image tweets and 1,605 (33.4%) non-visual Image tweets (Liu et al., 2007).

We use multimedia applications that make use of the content, Image, and social context of a graphic tweet to use machine learning (Khosla et al., 2014).

Text Features: The Chinese text is preprocessed by running every tweet through with a phrase segmenter, POS tagger, and Named Entity Recognizer (NER). We discovered that language is a good predictor of photo relationships: for example, tweets about a physical item and its color show a visual bias. Stop words and unusual words unlikely to reappear (freq 5) are removed from the last word function to make it more useful (Fattal, 2007; Hasselblad et al., 2007). The word characteristics are binary, representing just a word's existence (or absence). Another benefit is that we include the taught subject from LDA. We also track POS intensity (the ratio of nouns, verbs, adjectives, adverbs, and pronouns in a Twitter post) or the presence of various named entity classes. Because visual tweets reference tangible things, user names (e.g., the second row in Figure 4.3), locations, and items, these aspects are valuable. We additionally test for the existence of @ mentions, geographical locations, and URLs using 4th microblog-specific capabilities (Voigt et al., 2007).

Image Features: We avoid object recognition, which is typical in multimedia (TREC-MM) research since photos in Image twitter present a wide range of kinds. Facial recognition is used as an example, with the number of faces appearing being recorded. Faces are frequently the posters, friends, or relatives. We have implemented a composites founder feature that is enabled whenever a person's name and face are presented for the same purpose (Kilgarriff et al., 2014). Faces are spotted in 22.2% of the photos in our sample. Images with comparable content have a similar Image text relationship. To do so, we group the photographs based on their visual similarities. We initially extract SIFT descriptive terms from the photos as inputs, and tend to cluster them to shape visual words by constructing a hierarchical visual vernacular tree. The graphic vocabulary of the manuscript of Images as documents is then subjected to LDA in addition to learning k hidden topics. The photo topic task is then encoded as a particular feature as a result (Adams and Friedland, 2011).

Context Features: We realize that the time of the publishing time influences whether a post is visual or not, therefore we add it as a function. We track whether or not the equipment used to submit the Image tweets is mobile when individuals share what they've just seen (visual tweet) (e.g., desktops). Our collection is completed with social features. Visitors on Weibo may leave a comment on the article (Ashoor et al., 2013). In our sample, 46.3% of Image tweets had at least one remark, and 21.5% have been retweeted, compared to 33.7% and 16.2% for text-only tweets, respectively. As a feature, we utilize the number of replies and retweets normalized by the writer's user's account amount of supporters. We also observe that in visual tweets, the author responds to the post (typically in response to her reader's remarks), so we encode it as a function as well. Lastly, we employ the following ratio to distinguish ordinary users from celebrities and organizational accounts (i.e., (# followers)/(# followed)) (Lillesand et al., 2015).

4.5. EXPERIMENT

The Naive Bayes adoption in Weka is used in a 10 fold cross experiment. The 3 types of characteristics were concatenated into a single vector using linear concatenation. Simple correctness is not even an adequate assessment criterion due to the imbalanced distribution (66.6)% of photo tweets are visual) (Zhang et al., 2019). As a result, we provide the F1 score that has been macro-averaged, as we believe both classes are equally meaningful. The macro-F1 score for the majority baseline (all visual) is 0.40. To better understand the influence of each feature layer, we begin with the single best feature (words, F1 = 64.8) and assess the increase (loss) in F1 as we add each characteristic one at a time. Table 4.1 displays the findings. The second-most helpful attribute is POS frequency, which increases F1 by 4.9 (Chen et al., 2013; Ferguson et al., 2003). This characteristic is effective in identifying non-visual tweets with extensive use of function words as a snapshot of content (e.g., noun) and function (e.g., pronoun) word distribution (e.g., pure exclamations). Topic, named entity recognition, and microblog-specific text characteristics all result in a minor speed boost. The inclusion of the two Image characteristics improves the baseline by a little amount. However, not all of the suggested context elements are applicable. The inclusion of posting time, device, and follower ratio marginally improves the word baseline, but the other three don't. Our ultimate classifier (Row 14), which contains all of the baseline features, earns an F1 of 70.5 (Nowak and Greenfield, 2010).

Table 4.1. Experimental Results and Feature Analysis

Class	Features	Macro-F_1 (%)
Text	(1): Words Only (Baseline)	64.8
	(2): (1) + Microblog-specific	65.2
	(3): (1) + Named Entities	65.3
	(4): (1) + Topic	66.6
	(5): (1) + POS Density	69.7
Image	(6): (1) + Topic	65.4
	(7): (1) + Face	65.7
Context	(8): (1) + Retweets	60.9 (–)
	(9): (1) + Comments	64.5 (–)
	(10): (1) + Replied by Author	64.7 (–)
	(11): (1) + Device	64.9
	(12): (1) + Follower Ratio	64.9
	(13): (1) + Posting Time	65.0
All	(14): (1–7 + 11–13)	70.5

We looked into the incidents that were misclassified further. While words are the greatest distinguishing trait, the content on microblogs is rather brief (tweets on Weibo are limited to 140 characters) (Zoran, 1984). Because of the text's shortness, the word characteristic is sparse, providing little information to the classifier. Word characteristics are useless in extreme cases, such as (predecessor hall of the Wu family), when completion of the terms remains unusual or out of vocabulary (Feng et al., 2014). This helps explain why the words' base plateaus around 64.8 F1. Traditional natural language processing algorithms are particularly challenged by the informal speech employed in microblogs, such as neologisms and misspellings. We've seen a lot of cases where our word segment and isolated word recognition tools handle spelling errors improperly. One example is a graphic tweet that misspells (a cartoon character) as "J." This was not correctly identified as an identified entity by the NER tool. The eventual problem was caused by the transmission of this error downstream in our system (Pesaresi et al., 2007; Taubman and Marcellin, 2012).

Face detection imperfection is another cause of classification mistakes, in addition to text characteristics. This, we believe, is related to the nature of the Images used in visual tweets (e.g., low photo quality, photo collages). In addition, we see a flaw in our context features. We took a look at various users' feeds and Image tweets and discovered that they all had distinct ways of tweeting. Some people are more likely to publish non-visual tweets than visual tweets, while others are the opposite. This is not represented in our suggested context features, and we feel that characteristics that take into account user behavioral traits would be quite useful (Gupta et al., 2016; Tinland et al., 1997).

4.6. CONCLUSION

With the ability to incorporate photographs in micro-blog entries, the sociable Web 2.0 has embraced media. We looked at these Image tweets from a variety of angles, including visual, linguistic, and sociocultural contexts. We find that Images from photo tweets show a broader range of types of Images than photos from Image-sharing websites, and that Image tweets vary from writing tweets in areas of topic content and posting time. The graphically figure reveals a tweet, in which the focal point of the Twitter post is represented in both the Image and the text, complementing one another, which is an essential difference we make concerning Image tweets. Non-visual tweets, on the other hand, employ the Image to embellish the text in a non-essential way–i.e., to pique interest in reading a message.

We created an automatic classifier that used selected features, Images, and context data to obtain a macro F1 of 70.5, a 5.7% increase more than a text-only background. We made the identified corpus available to the public to test and benchmark against to stimulate future research on these areas. We've also discovered that non-visual Twitter is often related to emotions, and we'll categorize such emotionally relevant Image tweets in the coming. We also want to perform similar research on other channels (such as Twitter) and compare the results to those obtained on Weibo.

REFERENCES

1. Adams, S. M., & Friedland, C. J., (2011). A survey of unmanned aerial vehicle (UAV) usage for imagery collection in disaster research and management. In: *9th International Workshop on Remote Sensing for Disaster Response* (Vol. 8, pp. 1–8).
2. Ashoor, G., Syngelaki, A., Poon, L. C. Y., Rezende, J. C., & Nicolaides, K. H., (2013). Fetal fraction in maternal plasma cell-free DNA at 11–13 weeks' gestation: Relation to maternal and fetal characteristics. *Ultrasound in Obstetrics & Gynecology, 41*(1), 26–32.
3. Bruni, E., Tran, N. K., & Baroni, M., (2014). Multimodal distributional semantics. *Journal of Artificial Intelligence Research, 49*(1), 1–47.
4. Castillo, C., (2016). *Big Crisis Data: Social Media in Disasters and Time-Critical Situations* (4th edn., pp. 5–9). Cambridge University Press.
5. Chen, T., Lu, D., Kan, M. Y., & Cui, P., (2013). Understanding and classifying image tweets. In: *Proceedings of the 21st ACM international conference on Multimedia* (Vol. 1, pp. 781–784).
6. Cheng, G., Varanasi, P., Li, C., Liu, H., Melnichenko, Y. B., Simmons, B. A., & Singh, S., (2011). Transition of cellulose crystalline structure and surface morphology of biomass as a function of ionic liquid pretreatment and its relation to enzymatic hydrolysis. *Biomacromolecules, 12*(4), 933–941.
7. Daly, S., & Thom, J. A., (2016). Mining and classifying image posts on social media to analyze fires. In: *ISCRAM* (Vol. 1, pp. 1–14).
8. Fattal, R., (2007). Image upsampling via imposed edge statistics. In: *ACM SIGGRAPH 2007* papers (Vol. 1, pp. 95-es).
9. Feng, T., Hong, Z., Fu, Q., Ma, S., Jie, X., Wu, H., & Tong, X., (2014). Application and prospect of a high-resolution remote sensing and geo-information system in estimating earthquake casualties. *Natural Hazards and Earth System Sciences, 14*(8), 2165–2178.
10. Ferguson, S. C., Blane, A., Perros, P., McCrimmon, R. J., Best, J. J., Wardlaw, J., & Frier, B. M., (2003). Cognitive ability and brain structure in type 1 diabetes: Relation to microangiopathy and preceding severe hypoglycemia. *Diabetes, 52*(1), 149–156.
11. Friedman, K. E., Reichmann, S. K., Costanzo, P. R., Zelli, A., Ashmore, J. A., & Musante, G. J., (2005). Weight stigmatization and ideological

beliefs: Relation to psychological functioning in obese adults. *Obesity Research, 13*(5), 907–916.

12. Gupta, A., Vedaldi, A., & Zisserman, A., (2016). Synthetic data for text localization in natural images. In: *Proceedings of the IEEE Conference on Computer Vision and Pattern Recognition* (Vol. 1, pp. 2315–2324).
13. Hasselblad, V., Stough, W. G., Shah, M. R., Lokhnygina, Y., O'Connor, C. M., Califf, R. M., & Adams Jr, K. F., (2007). Relation between dose of loop diuretics and outcomes in a heart failure population: Results of the ESCAPE trial. *European Journal of Heart Failure, 9*(10), 1064–1069.
14. Hodosh, M., Young, P., & Hockenmaier, J., (2013). Framing image description as a ranking task: Data, models and evaluation metrics. *Journal of Artificial Intelligence Research, 47*(1), 853–899.
15. Imran, M., Castillo, C., Diaz, F., & Vieweg, S., (2015). Processing social media messages in mass emergency: A survey. *ACM Computing Surveys (CSUR), 47*(4), 1–38.
16. Imran, M., Castillo, C., Lucas, J., Meier, P., & Vieweg, S., (2014). AIDR: Artificial intelligence for disaster response. In: *Proceedings of the 23rd International Conference on World Wide Web (WWW'14)* (Vol. 1, pp. 159–162). Companion Seoul, Korea: ACM.
17. Imran, M., Elbassuoni, S., Castillo, C., Diaz, F., & Meier, P., (2013). Extracting information nuggets from disaster-Related messages in social media. *ISCRAM, 201*(3), 791–801.
18. Ishiguro, K., Kimura, A., & Takeuchi, K., (2012). Towards automatic image understanding and mining via social curation. In: *2012 IEEE 12th International Conference on Data Mining* (Vol. 1, pp. 906–911). IEEE.
19. Karpathy, A., Joulin, A., & Fei-Fei, L. F., (2014). Deep fragment embeddings for bidirectional image sentence mapping. *Advances in Neural Information Processing Systems, 27*(1), 2–8.
20. Kasturi, R., Goldgof, D., Soundararajan, P., Manohar, V., Garofolo, J., Bowers, R., & Zhang, J., (2008). Framework for performance evaluation of face, text, and vehicle detection and tracking in video: Data, metrics, and protocol. *IEEE Transactions on Pattern Analysis and Machine Intelligence, 31*(2), 319–336.

21. Katsaggelos, A. K., Biemond, J., Schafer, R. W., & Mersereau, R. M., (1991). A regularized iterative image restoration algorithm. *IEEE Transactions on Signal Processing, 39*(4), 914–929.

22. Kerle, N., Nex, F., Gerke, M., Duarte, D., & Vetrivel, A., (2019). UAV-based structural damage mapping: A review. *ISPRS International Journal of Geo-Information, 9*(1), 14.

23. Khosla, A., Das Sarma, A., & Hamid, R., (2014). What makes an image popular?. In: *Proceedings of the 23rd International Conference on World Wide Web* (Vol. 1, pp. 867–876).

24. Kilgarriff, A., Rychly, P., Smrz, P., & Tugwell, D., (2004). Itri-04-08 the sketch engine. *Information Technology, 105*(116), 105–116.

25. Lei, Y., Wang, Y., & Huang, J., (2011). Robust image hash in radon transform domain for authentication. *Signal Processing: Image Communication, 26*(6), 280–288.

26. Lillesand, T., Kiefer, R. W., & Chipman, J., (2015). *Remote Sensing and Image Interpretation* (2nd edn., pp. 2–8). John Wiley & Sons.

27. Liu, C. C., Liu, J. G., Lin, C. W., Wu, A. M., Liu, S. H., & Shieh, C. L., (2007). Image processing of FORMOSAT-2 data for monitoring the South Asia tsunami. *International Journal of Remote Sensing, 28*(13, 14), 3093–3111.

28. Nowak, D. J., & Greenfield, E. J., (2010). Evaluating the national land cover database tree canopy and impervious cover estimates across the conterminous United States: A comparison with photo-interpreted estimates. *Environmental Management, 46*(3), 378–390.

29. Ofli, F., Meier, P., Imran, M., Castillo, C., Tuia, D., Rey, N., & Joost, S., (2016). Combining human computing and machine learning to make sense of big (aerial) data for disaster response. *Big Data, 4*(1), 47–59.

30. Ozbulak, G., Aytar, Y., & Ekenel, H. K., (2016). How transferable are CNN-based features for age and gender classification?. In: *2016 International Conference of the Biometrics Special Interest Group (BIOSIG)* (Vol. 1, pp. 1–6). IEEE.

31. Pesaresi, M., Gerhardinger, A., & Haag, F., (2007). Rapid damage assessment of built-up structures using VHR satellite data in tsunami-affected areas. *International Journal of Remote Sensing, 28*(13, 14), 3013–3036.

32. Peters, R., & de Albuquerque, J. P., (2015). Investigating images as indicators for relevant social media messages in disaster management. In: *ISCRAM* (Vol. 1, pp. 5–9).
33. Rudra, K., Banerjee, S., Ganguly, N., Goyal, P., Imran, M., & Mitra, P., (2016). Summarizing situational tweets in crisis scenario. In: *Proceedings of the 27th ACM Conference on Hypertext and Social Media* (Vol. 1, pp. 137–147).
34. Russakovsky, O., Deng, J., Su, H., Krause, J., Satheesh, S., Ma, S., & Fei-Fei, L., (2015). ImageNet large scale visual recognition challenge. *International Journal of Computer Vision, 115*(3), 211–252.
35. Simonyan, K., & Zisserman, A., (2014). *Very Deep Convolutional Networks for Large-Scale Image Recognition* (2nd edn, pp. 3–7).
36. Starbird, K., Palen, L., Hughes, A. L., & Vieweg, S., (2010). Chatter on the red: What hazards threat reveals about the social life of microblogged information. In: *Proceedings of the 2010 ACM Conference on Computer Supported Cooperative Work* (Vol. 1, pp. 241–250).
37. Taubman, D., & Marcellin, M., (2012). *JPEG2000 Image Compression Fundamentals, Standards and Practice: Image Compression Fundamentals, Standards and Practice* (Vol. 642, pp. 2–9). Springer Science & Business Media.
38. Tien Nguyen, D., Alam, F., Ofli, F., & Imran, M., (2017). *Automatic Image Filtering on Social Networks Using Deep Learning and Perceptual Hashing During Crises* (2nd edn, pp. 2–9).
39. Tinland, B., Pluen, A., Sturm, J., & Weill, G., (1997). Persistence length of single-stranded DNA. *Macromolecules, 30*(19), 5763–5765.
40. Turker, M. U. S. T. A. F. A., & San, B. T., (2004). Detection of collapsed buildings caused by the 1999 Izmit, Turkey earthquake through digital analysis of post-event aerial photographs. *International Journal of Remote Sensing, 25*(21), 4701–4714.
41. Voigt, S., Kemper, T., Riedlinger, T., Kiefl, R., Scholte, K., & Mehl, H., (2007). Satellite image analysis for disaster and crisis-management support. *IEEE Transactions on Geoscience and Remote Sensing, 45*(6), 1520–1528.
42. Wang, Y., Li, Y., & Luo, J., (2016). Deciphering the 2016 us presidential campaign in the twitter sphere: A comparison of the trumpists and clintonists. In: *Tenth International AAAI Conference on Web and Social Media* (Vol. 1, pp. 2–9).

43. Yang, X., Tang, K., Zhang, H., & Cai, J., (2019). Auto-encoding scene graphs for image captioning. In: *Proceedings of the IEEE/CVF Conference on Computer Vision and Pattern Recognition* (Vol. 1, pp. 10685–10694).

44. Yosinski, J., Clune, J., Bengio, Y., & Lipson, H., (2014). How transferable are features in deep neural networks?. *Advances in Neural Information Processing Systems, 27*(2), 2–9.

45. Yu, L., Asur, S., & Huberman, B. A., (2011). *What Trends in Chinese Social Media* (2nd edn., pp. 2–6).

46. Zauner, C., (2010). *Implementation and Benchmarking of Perceptual Image Hash Functions* (Vol 1, pp. 5–10).

47. Zhang, C., Song, D., Huang, C., Swami, A., & Chawla, N. V., (2019). Heterogeneous graph neural network. In: *Proceedings of the 25th ACM SIGKDD International Conference on Knowledge Discovery & Data Mining* (Vol. 1, pp. 793–803).

48. Zhao, X., Zhu, F., Qian, W., & Zhou, A., (2013). Impact of multimedia in sina weibo: Popularity and life span. In: *Semantic Web and Web Science* (Vol. 1, pp. 55–65). Springer, New York, NY.

49. Zoran, G., (1984). Towards a theory of space in narrative. *Poetics Today, 5*(2), 309–335.

CHAPTER 5

USES OF DIGITAL IMAGE PROCESSING

CONTENTS

5.1. INTRODUCTION

Digital processing of image is the procedure of handling digital photos with the help of a computer. It is an area of signals and systems that focuses on visuals in particular. DIP emphasis on the advancement of a computer system capable of processing image. The system takes a digital image as input and processes it by using effective algorithms to generate an image as a result. Adobe Photoshop is the most prominent example. It's among the most extensively used applications for DIP (Billingsley, F. C. 1970; Erhardt, 2000).

As digital processing of image has so many applications and affects nearly all technical areas, only the most important DIP appl Digital processing of image isn't restricted to merely adjusting the spatial resolution of commonly obtained pictures. It isn't restricted to increasing the photo's brightness, etc. Electromagnetic waves are comparable to a stream of particles traveling at the speed of light. Every particle carries an energy bundle. This packet of energy is known as a photon (Pitas and Venetsanopoulos, 1992).

The electromagnetic spectrum is depicted below in terms of photon energy (Figure 5.1).

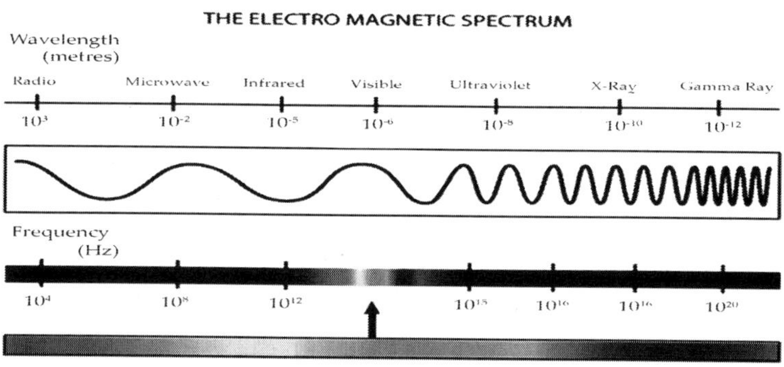

Figure 5.1. The EM (electromagnetic spectrum).

Source: https://earthsky.org/space/what-is-the-electromagnetic-spectrum/.

Only the visible spectrum is seen in the electromagnetic spectrum. Visible spectrum consists primarily of 7 hues that are popularly known as (VIBGOYR). The letters VIBGOYR represent the colors violet, indigo, blue, green, orange, yellow, and Red (Ravikumar and Arulmozhi, 2019).

However, this doesn't negate the presence of additional elements in the spectrum. The visible section, in which we observed all the things, is everything that our human sight can see. A camera, on the other hand, can see things that the naked eye can't. For instance, x rays, gamma rays, and so on. As a result, all of that information is analyzed using digital processing of image.

This discussion raises the question of why it is necessary to evaluate all the additional electromagnetic spectrum material. The appropriate response comes in the point that other technologies, like XRay, are extensively utilized in the medical industry. Gamma ray assessment is needed since it is frequently employed in nuclear medicine and astronomical monitoring. The same may be said for the remaining EM spectrum (Chu et al., 1985; Hijazi and Madhavan, 2008).

Digital processing of image is a collection of processes and approaches for manipulating pictures on a computer. DIP involves the implementation of several kinds of operations on pictures. Fundamentally, an image is a 2 D communication. The signal function is normally f(x, y), where the values of x and y at a certain place determine the pixel at that point. Image is just a 2D array of numbers ranging from 0 to 255. Image processing is influenced by a number of things. There are a number of reasons for image processing (Ekstrom, 2012; Fadnavis, 2014).

Processing of Image Assist in

- Enhancement in digital data stored.
- Automation of working with pictures.
- Improved optimization of picture leading to effective storage and communication.

Processing of image has advanced significantly over time, and there are numerous commercial uses of image processing today.

5.2. USES OF DIGITAL IMAGE PROCESSING (DIP)

The following are some of the primary areas in which digital processing of image is commonly utilized.

a. Image sharpening and restoration.
b. Remote sensing
c. Medical field
d. Transmission and encoding
e. Processing of color
f. Machine vision
g. Processing of video

h. Pattern identification
i. Microscopic Imaging
j. Others

5.2.1. Image Sharpening and Restoration

This method refers to the enhancement of photos acquired by a modern camera or the manipulation of such pictures to achieve a desired effect. It refers to the standard functionality of Photoshop. Zooming, fading, sharpening, gray scale to conversion into colored image, identifying edges and conversely, retrieval of image, and image identification are all examples of this (Lin et al., 2008; Ouyang et al., 2010). The following are some common instances.

The original picture is shown in Figure 5.2.

Figure 5.2. Original picture.

Source: https://www.weforum.org/agenda/2015/11/why-is-einstein-famous/.

And the zoomed picture is shown in Figure 5.3.

Figure 5.3. The zoomed picture.

Source: https://www.weforum.org/agenda/2015/11/why-is-einstein-famous/.

The Blur picture is shown in Figure 5.4.

Figure 5.4. The blur picture.

Source: https://www.weforum.org/agenda/2015/11/why-is-einstein-famous/.

The sharp picture is shown in Figure 5.5.

Figure 5.5. The sharp picture.

Source: https://www.weforum.org/agenda/2015/11/why-is-einstein-famous/.

The edges is shown in Figure 5.6.

Figure 5.6. Picture with edges.

Source: https://www.weforum.org/agenda/2015/11/why-is-einstein-famous/.

5.2.2. Medical Field

The common uses of digital processing of image in this area are:

- Gamma ray imaging.
- X Ray Imaging
- PET scan
- Ultraviolet imaging
- Medical CT

In the area of medical, processing of image is utilized for several activities, including PET scan, X-ray imaging, medical CT, ultraviolet imaging, cancer cell processing of image, and a great deal more. Incorporation of processing of image to the realm of medical technology has significantly boosted the diagnostic procedure (Saxton et al., 1979; McAndrew, 2016).

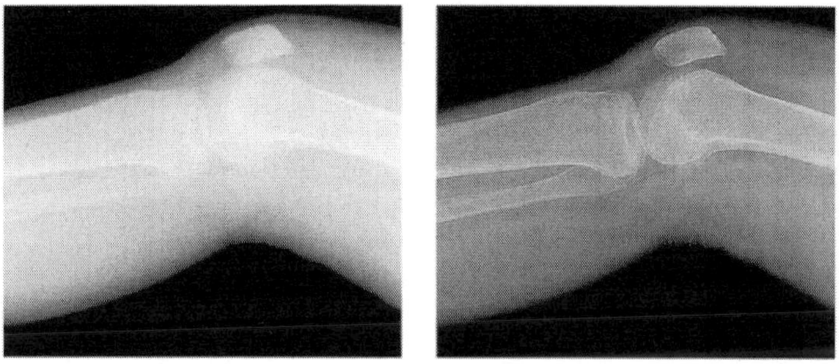

Figure 5.7. Digital picture of bone.

Source:https://www.intechopen.com/chapters/67331.

The original picture is on left side. The processed picture is on right side. can noticed that the picture after processing is much better and can generally be utilized for more accurate diagnoses (Figure 5.7).

5.2.3. Ultra Violet Imaging

In the area of remote sensing, the region of earth is examined with the help of a satellite or from a relatively high vantage point and then evaluated for information. One use of DIP in the realm of remote sensing is the detection of earthquake-related damage to infrastructure. Since it takes longer to comprehend damages, even when the attention is on severe damages. Because the area affected by an earthquake can be so vast, it isn't always feasible to inspect it visually in order to quantify the damage. Even if true, it is a tedious and time-consuming technique. Digital processing of image provides a solution to the situation. The picture of affected area is taken from above and then processed to examine the different forms of earthquake-caused damage (Maerz, 1998; Kheradmand and Milanfar, 2014) (Figure 5.8).

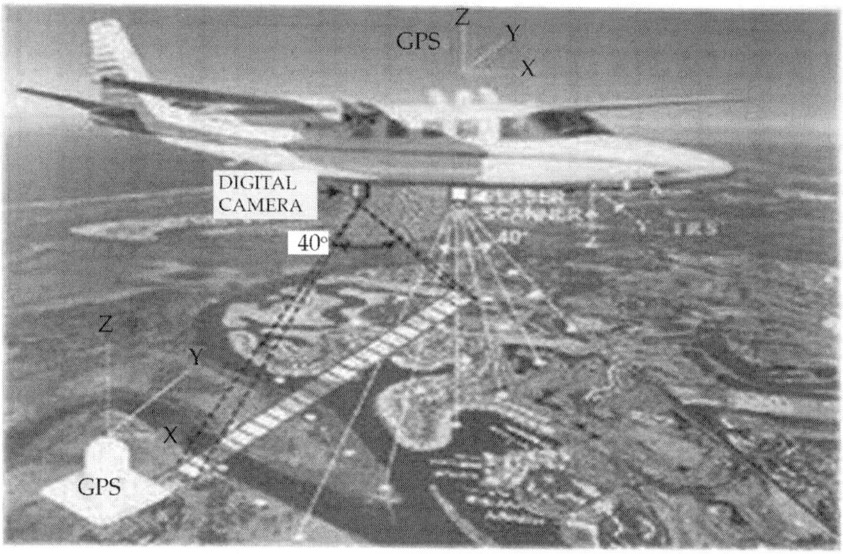

Figure 5.8. Application in ultraviolet imaging.

Source; https://www.tutorialspoint.com/dip/applications_and_usage.htm.

The analysis comprises of the following main steps:

- The removal of edges;
- Examination and improvement of several kinds of edges.

5.2.4. Transmission and Encoding

The 1st picture transferred over the wire was via an underwater connection from London (U.K) to New York (USA). Imagine being capable to witness live video feeds or footage of CCTV from one continent to the other with only a few seconds of delay. It implies that a significant amount of progress has been made in this area as well. This field is concerned with both transmission and encoding. Several distinct formats have been created to encode photographs for low or high bandwidth and stream them over the internet or via other means (Lin et al., 2021).

5.2.5. Machine Vision

Among several hurdles that robots confront today, increasing their vision remains one of the greatest obstacles. Make the robot capable of seeing, identifying, and identifying obstacles, etc. This area has contributed a substantial amount of work, and an entirely new discipline of object recognition has been formed to work on it (Meng et al., 2020).

5.2.6. Detection of Hurdle

This is a general image processing issue that involves recognizing different objects in a picture and then estimating the distance among the robot and the hurdles.

The bulk of today's robots operate by following a line, and so are referred to as line follower robots. This permits a robot to follow its pathway and complete jobs. Processing of image was used to accomplish this (Dobeš et al., 2010; Chen et al., 2015) (Figure 5.9).

Figure 5.9. Detection of hurdle.

Source: https://www.tutorialspoint.com/dip/applications_and_usage.htm.

5.2.7. Processing of Color

It involves the processing of colored pictures and the utilization of various color spaces. For instance RGB color model and HSV. In addition, communication, storage, and encoding of the color pictures are investigated (Andrews, 1974).

5.2.8. Recognition of Pattern

It encompasses research from processing of image and other disciplines, such as machine learning. In pattern identification, processing of image is used to recognize the objects in a picture, followed by the application of machine learning in order to educate the system for variations in pattern. Pattern identification is utilized in computer-assisted diagnostics, handwriting identification, image identification, etc. (Chen and Pock, 2016).

5.2.9. Processing of Video

Video is nothing more than the rapid movement of images. The video's quality is determined by the total number of frames and the quality of every frame. Reduction of noise, detail improvement, Detection of motion, conversion of frame rate, conversion of aspect ratio, conversion of color space, and so on are all part of video handling. Video is just a fast-moving series of images. Video Processing employs a variety of image processing methods. Removal of noise, stabilization of picture, conversion of frame rate, detail improvement, and other video processing techniques are only a few examples (Penczek, P. A. 2010; Zhu and Milanfar, 2011).

5.2.9.1. Beginning with Processing of Image in Python

Let's begin with some fundamental Image-associted jobs in Python. The Python Imaging Library is utilized for a variety of image processing jobs.

5.2.9.2. Installing Python

```
pip install pillow
```

We can go on to the code now that PIL is installed.

We begin by utilizing several matplotlib methods.

```
import matplotlib.image as pic
import matplotlib.pyplot as plt
import numpy as np
%matplotlib inline
```

The image below will be read. image1.jpg is the name of the file.

Source:https://www.analyticsvidhya.com/blog/2021/05/digital-image-processing-real-life-applications-and-getting-started-in-python/

```
# reading jpg pic
pic = pic.imread('pic1.jpg')
plt.imshow(pic)
```

```
<matplotlib.image.AxesImage at 0x1e6d5ee3bc8>
```

Source:https://www.analyticsvidhya.com/blog/2021/05/digital-image-processing-real-life-applications-and-getting-started-in-python/.

The picture is read.

```
# transforming the image's shape
Rum1 = pic[:, :, 0]
plt.imshow(Rum1)
```

Now the shape of picture is altered.

```
<matplotlib.image.AxesImage at 0x1e6d5f9da08>
```

We will now convert it to the "hot" colormap. To learn more about colormap, click on the link provided.

plt.imshow(Rum1, cmap ='hot')

plt.colorbar()

Image looks like:

```
<matplotlib.colorbar.Colorbar at 0x1e6d6053d48>
```

Now a diverse colormap is tried.

picplot = plt.imshow(Rum1)

picplot.set_cmap('nipy_spectral')

Output of picture:

The purpose for utilizing colormaps is because having a consistent colormap can help in a variety of applications. Learn about Colormaps: Selecting Colormaps in Matplotlib.

Let's have a glance at why a photo is referred to as a two-dimensional array.

#data type of Rum1

print(typc(Rum1))

Output: <class 'numpy.ndarray'>

print(Rum1)

[[92 91 89 … 169 168 169] [110 110 110 … 168 166 167] [100 103 108 … 164 163 164]

…

[97 96 95 … 144 147 147] [99 99 98 … 145 139 138] [102 102 103 … 149 137 137]]

The purpose of the dots is to indicate that there are several more points of data between them. Undoubtedly, however, it is all numerical data (Ramesh and Nityananda, 1986; Fedosov et al., 2019).

Determine the length of an array.

len(Rum1)

Output: 320

len(Rum1[300])

Output: 658

This tells us the image's dimensions and amount of pixels: 320*658.

We'll double-check this later.

We collaborate with PIL.

from PIL import Pic

This picture file will be used, titled as: people.jpg.

pic2 = Image.open('people.jpg')
plt.imshow(pic2)

```
<matplotlib.image.AxesImage at 0x1e6d614bf88>
```

The picture is read.

The photo is resized now.

pic2.thumbnail((50, 50), Image. ANTIALIAS) # resizes photo in-place

picplot = plt.imshow(pic2)

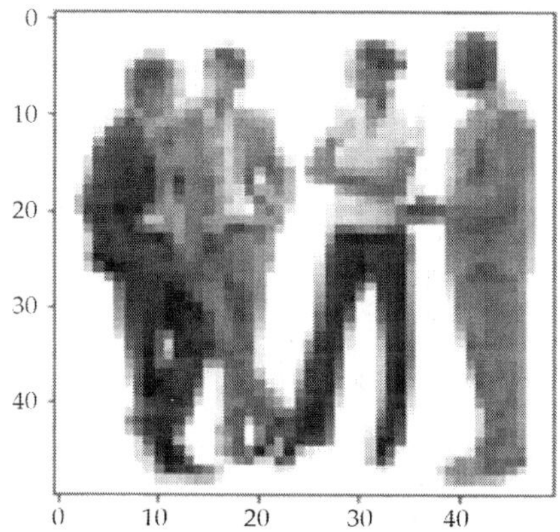

picplot1 = plt.imshow(pic2, interpolation="nearest")

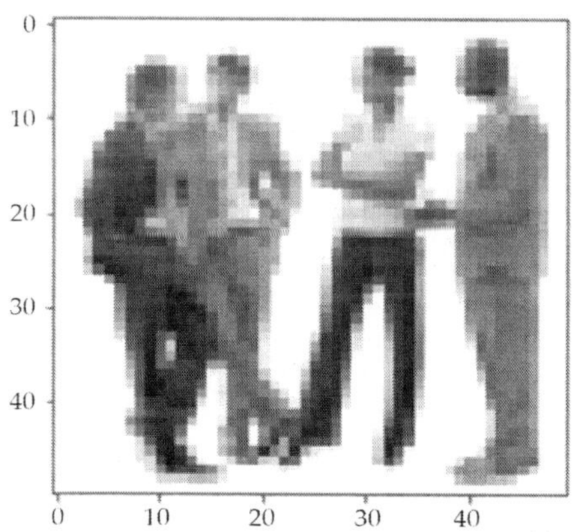

picplot2 = plt.imshow(pic2, interpolation="bicubic")

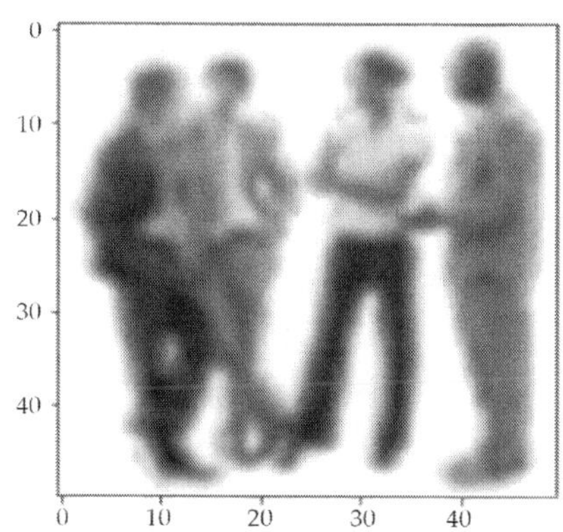

But why will be photos blurred on purpose in Processing of Image? Well, it becomes challenging for Classification of Pattern and Computer

Vision algorithms to process pictures that are exceptionally sharp. Therefore, photos are blurred to make them smooth. Additionally, blurring makes the color shift from one side of an image to the other much smoother (Rani et al., 2016; Mohan et al., 2022).

Now, let's verify the image's dimensions that was worked on before.

```
#some more stimulating things
file='pic1.jpg'
with Image.open(file) as photo:
width, height = pic.size
#Image width, height is be attained
```

```
In [16]: width
Out[16]: 658

In [17]: height
Out[17]: 320
```

These are the same measurements we obtained earlier. As a result, it can be deduced that the photo is 320*658 pixels.

Try flipping and transposing the photo as well.

```
#Comparative Path
pic3 = Image.open("pic1.jpg")
#Angle provided
pic_rot= pic3.rotate(180)
#Saved in same comparative location
pic_rot.save("rotated_picture.jpg")
```

This is the rotated photo

```
#transposing photo
transposed_pic = pic3.transpose(Image. FLIP_LEFT_RIGHT)
#Saved in the same comparative location
transposed_pic.save("transposed_pic.jpg")
```

This is the transposed photo.

This illustration will help us comprehend the idea of dimension.

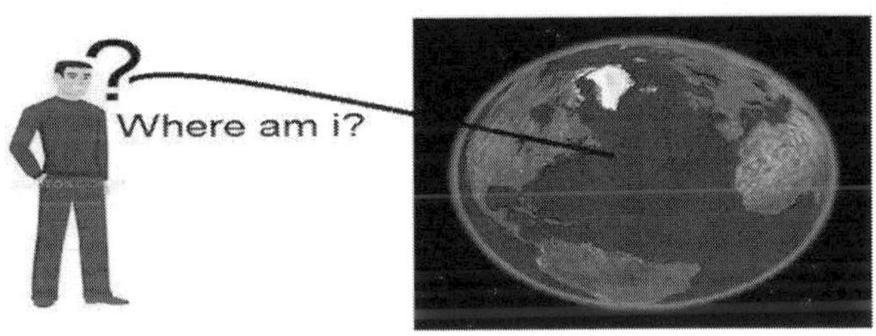

Consider the case where your buddy lives on the moon and want to give you a gift for your birthday. He inquiries about your earthly dwelling. The main issue is that the lunar courier service doesn't recognize alphabetical addresses and only recognizes numerical coordinates. So, how do you tell him where you are on the planet? (Park et al., 1994; Yao et al., 2006)

This is where the idea of dimensions originates. Dimensions describe the minimal quantity of points needed to locate a specific object within the given volume. So let's return to our original instance in which you must convey your location on earth to this buddy on the moon. You send him three coordinate pairs. The 1st is known as the longitude, the 2nd as the latitude, and the 3rd as the altitude (Buffington et al., 1976; Debayle et al., 2006).

These three coordinates establish your location on the globe. The 1st two determine your location, while the 3rd indicates your elevation above sea level. As a result, you only need three coordinates to determine your location on the planet. This means you live in a 3D. As a result, this not just solves the query of dimension, but also the query of why we exist in a 3D world. Because we're studying this subject in the context of digital processing of image, we'll use a picture to illustrate the idea of dimension (Wang, 2017; Zhang et al., 2019).

5.2.10. Dimensions of Picture

Now, if we inhabit a 3D environment, what are the dimensions of the images we capture? Since a picture is 2D, it is likewise characterized as a 2D signal. The only dimensions of picture are its height and width. A photograph doesn't have depth. Please observe the Figure 5.10.

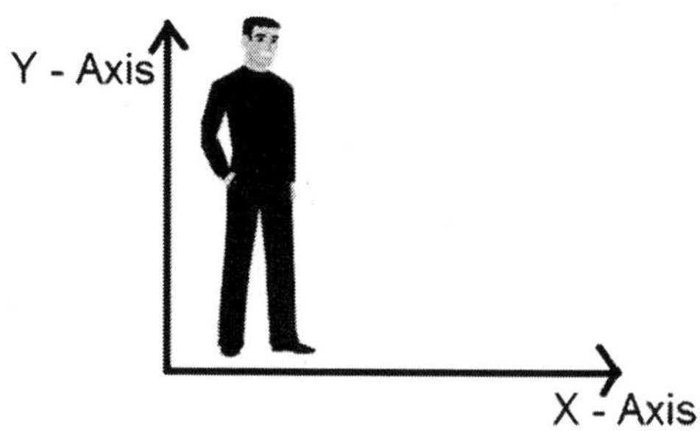

Figure 5.10. Dimensions of picture.

Source: https://www.tutorialspoint.com/dip/applications_and_usage.htm.

If you examine the Figure 5.10, you will notice that it has just two axes: the height and width. This photograph lacks the ability to convey depth

(Barbu and Favini, 2014). The image is called a 2D signal for this reason. But human eye is capable of perceiving 3D things; this will be discussed in greater detail in the subsequent course on how the camera functions and how images are received (Yao et al., 2007; Wang et al., 2020).

This topic raises the question of how 3D systems are created from two-dimensional (2D) ones. rom the picture above it can be seen that it is a 2D image. One more dimension is required to convert it to three dimensions (Chao et al., 2006). Let's consider time as the 3rd dimension; in that circumstance, we'll move this 2D photo over time. The same idea that assist us comprehend the depth of distinct items on a screen in television. Does this imply that what we see on television or on our television screens is 3D? Yes, we can do it (Chang et al., 2018; He et al., 2019).

Because, in the instance of television, we are playing a video. A video consists solely of 2D images moving in time dimension. As 2D objects move through the 3rd dimension, time, it can be claimed that it is 3D (Molina et al., 2001; Papyan and Elad, 2015).

5.3. DIVERSE DIMENSIONS OF SIGNALS

5.3.1. One-Dimensional Signal

The typical instance of a one-dimensional signal is the waveform. Mathematically it can be symbolized as

A(x) = waveform

where x→ independent variable.

As it is a one-dimensional signal, so just 1 variable/parameter x is used.

Pictorial illustration of a 1D signal is shown in Figure 5.11.

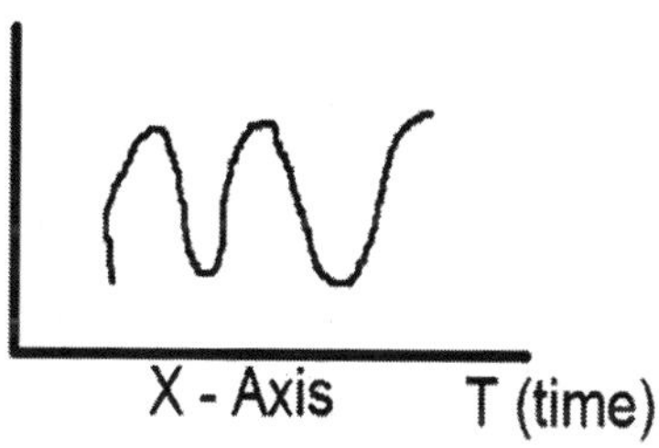

Figure 5.11. Pictorial illustration of a 1D signal.

Source: https://www.tutorialspoint.com/dip/applications_and_usage.htm.

Figure 5.11 displays a 1D signal. This leads to the next point, which is, if it is a 1D signal, why does it possesses 2 axes? The solution to this query is that, while being a 1D signal, it is depicted in a 2D space. Alternatively, it might be said that the space used to express this signal is 2D. Therefore, it appears to be a 2D signal (Blomgren et al., 1997; Horisaki et al., 2009).

Looking at the diagram below may help you better comprehend the notion of 1D.

Contemplate the figure above as a real line with positive (+ve) values from one point to another, referring back to the original talk on dimension. Now, if we need to describe the position of any point on this particular line, we simply need one integer, or dimension (George et al., 2012).

5.3.2. Two-Dimensional Signal

The common instance of the 2D signal is a picture, which has been discussed already above.

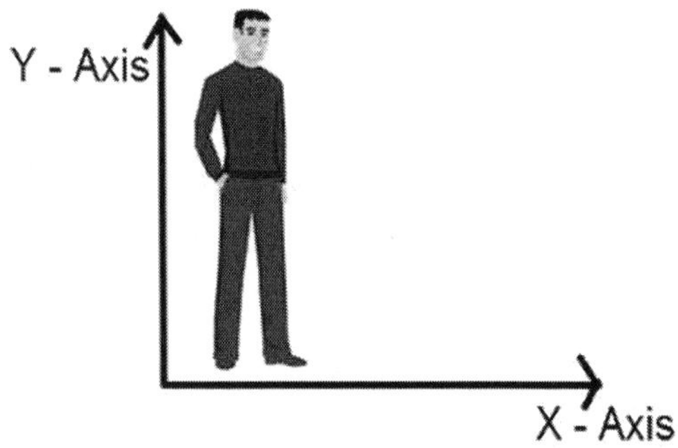

Figure 5.12. The common instance of a 2D signal.

Source: https://www.tutorialspoint.com/dip/applications_and_usage.htm.

As noticed already that the picture is 2D signal (Figure 2D), for example: it possesses two dimensions. Mathematically it can be expressed as:

A (x, y) = Image

where x & y→ two variables.

The notion of two dimension can be described in mathematics as:

Label the four corners of a square in the above diagram with the letters A, B, C, and D, correspondingly. It can be observed that these two parallel segments in the picture AB and CD link up to form a square if one of them is called AB and the second one CD. Since every line segment represents a single dimension, these two segments of line represent two dimensions (Wallmüller, 2007; Kanaev et al., 2012).

5.3.3. Three-Dimensional Signal

3D signal, as its name suggests, refers to signals with three dimensions. The most prevalent example has already been covered, which is our planet. We inhabit a world with three dimensions. This instance has been carefully discussed. One other instance of a 3 D signal is a cube or volumetric data, with 3D cartoon characters being the most prevalent (Brown and Seales, 2004; Li and Yang, 2010).

Mathematically it can be expressed as:

A(x, y, z) = animated character.

One more axis or dimension Z is indulged in three dimension, that provides the delusion of depth. In the Cartesian co-ordinate system 3D signal can be observed as (Figure 5.13):

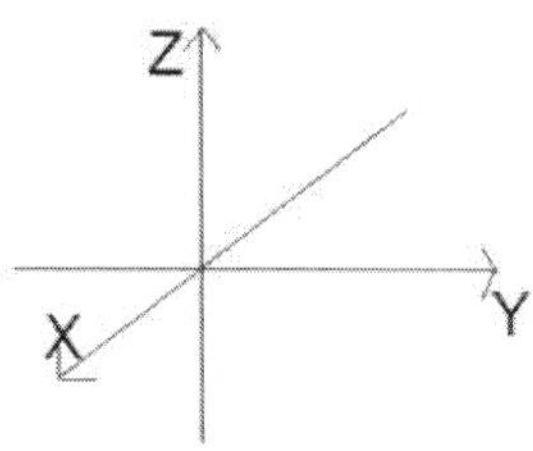

Figure 5.13. 3D signals.

Source: https://www.tutorialspoint.com/dip/applications_and_usage.htm.

5.3.4. Four-Dimensional Signal

Four dimensions are included in a 4D signal. The 1st three are identical to those of the 3D signal (X, Y, and Z), and the 4th is T (time), which is added to them. The temporal dimension is a means to measure variation and is typically referred to as time (Noll, 1997; Ingole and Shandilya, 2006). A 4D signal can be expressed mathematically as:

A(x, y, z, t) = animated movie.

The typical instance of a 4D signal can generally be an animated 3D movie. As every character is the 3D character and they are progressed with respect to time, because of which an impression of a 3D movie is seen just like the real world (de Polo, 2002).

REFERENCES

1. Andrews, H. C., (1974). Digital image restoration: A survey. *Computer, 7*(5), 36–45.
2. Barbu, T., & Favini, A., (2014). Rigorous mathematical investigation of a nonlinear anisotropic diffusion-based image restoration model. *Electronic Journal of Differential Equations, 129*(2014), 1–9.
3. Billingsley, F. C., (1970). Applications of digital image processing. *Applied Optics, 9*(2), 289–299.
4. Blomgren, P., Chan, T. F., Mulet, P., & Wong, C. K., (1997). Total variation image restoration: Numerical methods and extensions. In: *Proceedings of International Conference on Image Processing* (Vol. 3, pp. 384–387). IEEE.
5. Brown, M. S., & Seales, W. B., (2004). Image restoration of arbitrarily warped documents. *IEEE Transactions on Pattern Analysis and Machine Intelligence, 26*(10), 1295–1306.
6. Buffington, A., Crawford, F. S., Muller, R. A., Schwemin, A. J., & Smits, R. G., (1976). Active image restoration with a flexible mirror. In: *Imaging Through the Atmosphere* (Vol. 75, pp. 90–96). SPIE.
7. Chang, Y., Yan, L., Fang, H., Zhong, S., & Liao, W., (2018). HSI-DeNet: Hyperspectral image restoration via convolutional neural network. *IEEE Transactions on Geoscience and Remote Sensing, 57*(2), 667–682.
8. Chao, S. M., & Tsai, D. M., (2006). Astronomical image restoration using an improved anisotropic diffusion. *Pattern Recognition Letters, 27*(5), 335–344.
9. Chen, Y., & Pock, T., (2016). Trainable nonlinear reaction diffusion: A flexible framework for fast and effective image restoration. *IEEE Transactions on Pattern Analysis and Machine Intelligence, 39*(6), 1256–1272.
10. Chen, Y., Yu, W., & Pock, T., (2015). On learning optimized reaction diffusion processes for effective image restoration. In: *Proceedings of the IEEE Conference on Computer Vision and Pattern Recognition* (pp. 5261–5269).
11. Chu, T. C., Ranson, W. F., & Sutton, M. A., (1985). Applications of digital-image-correlation techniques to experimental mechanics. *Experimental Mechanics, 25*(3), 232–244.

12. de Polo, A., (2002). Digital picture restoration and enhancement for quality archiving. In: *2002 14th International Conference on Digital Signal Processing Proceedings. DSP 2002 (Cat. No. 02TH8628)* (Vol. 1, pp. 99–102). IEEE.
13. Debayle, J., Gavet, Y., & Pinoli, J. C., (2006). General adaptive neighborhood image restoration, enhancement and segmentation. In: *International Conference Image Analysis and Recognition* (pp. 29–40). Springer, Berlin, Heidelberg.
14. Dobeš, M., Machala, L., & Fürst, T., (2010). Blurred image restoration: A fast method of finding the motion length and angle. *Digital Signal Processing, 20*(6), 1677–1686.
15. Ekstrom, M. P., (2012). *Digital Image Processing Techniques* (Vol. 2). Academic Press.
16. Erhardt-Ferron, A., (2000). *Theory and Applications of Digital Image Processing* (Vol. 1, No. 2, p. 3). University of Applied Sciences Offenburg.
17. Fadnavis, S., (2014). Image interpolation techniques in digital image processing: An overview. *International Journal of Engineering Research and Applications, 4*(10), 70–73.
18. Fedosov, V. P., Ibadov, R. R., Ibadov, S. R., & Voronin, V. V., (2019). Restoration of the blind zone of the image of the underlying surface for radar systems with doppler beam sharpening. In: *2019 Radiation and Scattering of Electromagnetic Waves (RSEMW)* (pp. 424–427). IEEE.
19. George, A., Rajakumar, B. R., & Suresh, B. S., (2012). Markov random field based image restoration with aid of local and global features. *International Journal of Computer Applications, 48*(8).
20. He, J., Dong, C., & Qiao, Y., (2019). Modulating image restoration with continual levels via adaptive feature modification layers. In: *Proceedings of the IEEE/CVF Conference on Computer Vision and Pattern Recognition* (pp. 11056–11064).
21. Hijazi, A., & Madhavan, V., (2008). A novel ultra-high speed camera for digital image processing applications. *Measurement Science and Technology, 19*(8), 085503.
22. Horisaki, R., Nakao, Y., Toyoda, T., Kagawa, K., Masaki, Y., & Tanida, J., (2009). A thin and compact compound-eye imaging system incorporated with an image restoration considering color shift, brightness variation, and defocus. *Optical Review, 16*(3), 241–246.

23. Ingole, K. R., & Shandilya, V. K., (2006). Image restoration of historical manuscripts. *Int. J. Comput. Sci. Eng. Technol., 2*(4), 2229–3345.
24. Kanaev, A. V., Smith, L. N., Hou, W. W., & Woods, S., (2012). Restoration of turbulence degraded underwater images. *Optical Engineering, 51*(5), 057007.
25. Kheradmand, A., & Milanfar, P., (2014). A general framework for regularized, similarity-based image restoration. *IEEE Transactions on Image Processing, 23*(12), 5136–5151.
26. Li, S., & Yang, B., (2010). A new pan-sharpening method using a compressed sensing technique. *IEEE Transactions on Geoscience and Remote Sensing, 49*(2), 738–746.
27. Lin, H., Si, J., & Abousleman, G. P., (2008). Orthogonal rotation-invariant moments for digital image processing. *IEEE Transactions on Image Processing, 17*(3), 272–282.
28. Lin, S., Chi, K., Wei, T., & Tao, Z., (2021). Underwater image sharpening based on structure restoration and texture enhancement. *Applied Optics, 60*(15), 4443–4454.
29. Maerz, N. H., (1998). Aggregate sizing and shape determination using digital image processing. In: *Center For Aggregates Research (ICAR) Sixth Annual Symposium Proceedings* (pp. 195–203).
30. McAndrew, A., (2016). *A Computational Introduction to Digital Image Processing* (Vol. 2). Boca Raton: CRC Press.
31. Meng, H., Yan, Y., Cai, C., Qiao, R., & Wang, F., (2020). A hybrid algorithm for underwater image restoration based on color correction and image sharpening. *Multimedia Systems,* 1–11.
32. Mohan, A., Dwivedi, R., & Kumar, B., (2022). Image restoration of landslide photographs using SRCNN. *Recent Trends in Electronics and Communication*, 1249–1259.
33. Molina, R., Núñez, J., Cortijo, F. J., & Mateos, J., (2001). Image restoration in astronomy: A Bayesian perspective. *IEEE Signal Processing Magazine, 18*(2), 11–29.
34. Noll, D., (1997). Variational methods in image restoration. In: *Recent Advances in Optimization* (pp. 229–245). Springer, Berlin, Heidelberg.
35. Ouyang, A., Luo, C., & Zhou, C., (2010). Surface distresses detection of pavement based on digital image processing. In: *International Conference on Computer and Computing Technologies in Agriculture* (pp. 368–375). Springer, Berlin, Heidelberg.

36. Papyan, V., & Elad, M., (2015). Multi-scale patch-based image restoration. *IEEE Transactions on Image Processing, 25*(1), 249–261.
37. Park, S. K., & Hazra, R., (1994). Image restoration versus aliased noise enhancement. In: *Visual Information Processing III* (Vol. 2239, pp. 52–62). SPIE.
38. Penczek, P. A., (2010). Image restoration in cryo-electron microscopy. In: *Methods in Enzymology* (Vol. 482, pp. 35–72). Academic Press.
39. Pitas, I., & Venetsanopoulos, A. N., (1992). Order statistics in digital image processing. *Proceedings of the IEEE, 80*(12), 1893–1921.
40. Ramesh, N., & Nityananda, R., (1986). Maximum entropy image restoration in astronomy. *Annual Review of Astronomy and Astrophysics, 24,* 127–170.
41. Rani, S., Jindal, S., & Kaur, B., (2016). A brief review on image restoration techniques. *International Journal of Computer Applications, 150*(12), 30–33.
42. Ravikumar, R., & Arulmozhi, V., (2019). Digital image processing-a quick review. *International Journal of Intelligent Computing and Technology (IJICT), 2*(2), 11–19.
43. Saxton, W. O., Pitt, T., & Horner, M., (1979). Digital image processing: The semper system. *Ultramicroscopy, 4*(3), 343–353.
44. Wallmüller, J., (2007). Criteria for the use of digital technology in moving image restoration. *The Moving Image, 7*(1), 78–91.
45. Wang, F., (2017). A study of digital image enhancement for cultural relic restoration. *International Journal of Engineering and Technical Research, 7*(11).
46. Wang, L., Kim, T. K., & Yoon, K. J., (2020). EventSR: From asynchronous events to image reconstruction, restoration, and super-resolution via end-to-end adversarial learning. In: *Proceedings of the IEEE/CVF Conference on Computer Vision and Pattern Recognition* (pp. 8315–8325).
47. Yao, Y., Abidi, B. R., & Abidi, M. A., (2007). Extreme zoom surveillance: System design and image restoration. *J. Multim., 2*(1), 20–31.
48. Yao, Y., Abidi, B., & Abidi, M., (2006). Digital imaging with extreme zoom: System design and image restoration. In: *Fourth IEEE International Conference on Computer Vision Systems (ICVS'06)* (pp. 52). IEEE.

49. Zhang, W., Dong, L., Pan, X., Zou, P., Qin, L., & Xu, W., (2019). A survey of restoration and enhancement for underwater images. *IEEE Access, 7*, 182259–182279.
50. Zhu, X., & Milanfar, P., (2011). Restoration for weakly blurred and strongly noisy images. In: *2011 IEEE Workshop on Applications of Computer Vision (WACV)* (pp. 103–109). IEEE.
51. Zhuang, P., & Ding, X., (2019). Divide-and-conquer framework for image restoration and enhancement. *Engineering Applications of Artificial Intelligence, 85*, 830–844.

CHAPTER 6

SYNTHESIZED SOUND QUALITIES AND SELF-PRESENTATION

CONTENTS

6.1. INTRODUCTION

Currently, voice interfaces have become increasingly popular, allowing for accessible interactions, intuitive, and hands-free (Addington, 1968). With the extensive use of voice assistants like Amazon's Alexa, Apple's Siri, Google's Assistant, and Microsoft's Cortana, a growing number of online and mobile apps have begun to integrate voice communication (Aronovitch, 1976).

Simultaneously, researchers have started to look at the expressive aspects of synthetic voices, moving far beyond the mechanics of voice recognition, voice synthesis, and conversation interpretation. The focus is shifting away from what the voice must say and toward how it might say. Users' choices for synthetic speeches when engaging with a speech agent or consuming material were the subject of this study (such as hearing audiobooks or following one-by-one directions). Several studies show that listeners choose voices that have comparable or complementary features to their own, like gender and personality (Black and Tokuda, 2005). Rather than, we look at a similar but separate subject in this paper: how do people should have their material sound? A feeling of identity is especially important on social networking platforms. With exception of visual elements like colors, fonts, and photos, social media consumers presently have little to no influence over how its material is generated for voice commands (Boyd, 2008). Rather, the sound has been governed by preferences in voice assistants or screen reader apps, in which the preset voices are typically lacking in diversification and emotion. We anticipate that, as voice assistants begin to give audio-based access to a wide variety of web content, not only will website and voice application developers be able to choose from a variety of synthesized voices, but these end-users will also be able to customize voices for their information in the form of a general "profile voice" or more information styling (Boyd, 2006) (Figure 6.1).

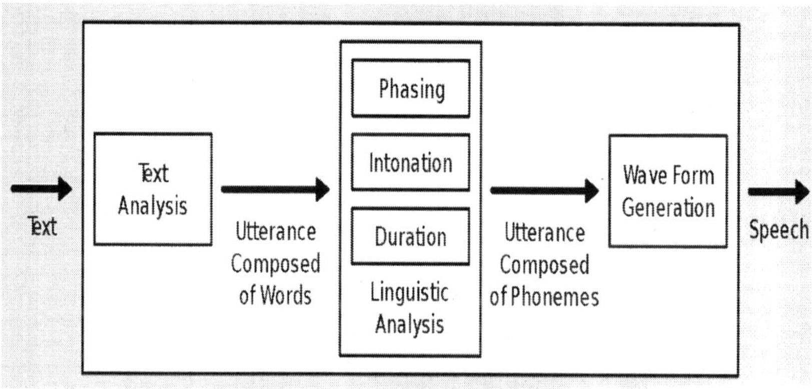

Figure 6.1. A typical TTS system is depicted in this diagram.

Source: https://en.wikipedia.org/wiki/Speech_synthesis.

We did semi-structured interview research with fifteen respondents to determine how social media consumers will like their material to sound as well as how they imagine listeners will react to that sound. Many respondents had been frequent users of social media and had prior voice interface expertise. Whereas the majority were young adults, their genders and nationalities were diverse (Braun et al., 2019). To encourage respondents to evaluate a broad range of emotional synthesized voice features, we presented every respondent using audio excerpts of their social media profile uttered by synthesized voices which diversified in accent, age, emotion, and gender. Respondents were asked how they thought about the offered voices for delivering their material and what factors they will consider if they wished to create an ideal customized voice (Cambre et al., 2020).

Our studies identified requirements for synthetic voices to effectively deliver social media material, like expressiveness, adaptability, and context-appropriateness. Respondents desired an accurate and consistent synthetic voice portrayal of themselves, as well as the option to express the emotion of certain postings and to select "fun" or "formal" voices for various platforms (such as LinkedIn versus Twitter). We also highlighted significant issues associated with voice customization, including the likelihood that accented synthetic speech might perpetuate prejudices (Cambre and Kulkarni, 2019) (Figure 6.2).

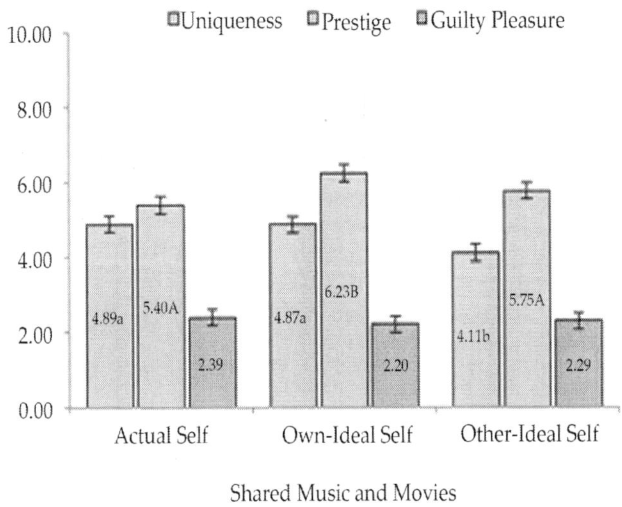

Figure 6.2. Graphical depiction of self-presentation.

Source: https://www.sciencedirect.com/science/article/abs/pii/S0747563218300141.

The following contributions are made by this chapter (Chou and Edge, 2012):

- a description of one's preferences for presenting one's socially-oriented material (as opposed to listening to voice-based information);
- The relationship between synthetic voice preferences and online impression management theory is empirically supported and interpreted.
- a list of technological, usability, and ethical issues, as well as design concerns for future work on socially located, self-customized speech synthesis systems.

6.2. SYNTHESIZED VOICE CHOICE

What characteristics distinguish a beautiful voice from a speech interaction? Listeners are drawn to voices that have comparable "personal" traits to their own according to studies on speech synthesis. Lee et al. discovered that a user's perception of social existence was enhanced when the voice's personality matched the user's personality in an experiment wherein participants had been required to listen to book reviews in synthetic voices (Lee et al., 2012). Braun et al. found that customizing the voice assistant's personality for every consumer resulted in better levels of favorability and trust than using a preset personality (Braun et al., 1989). Whenever the accent and perceived voice gender matched their own or the material being stated, people appeared to judge the speech more highly (Clark et al., 2019) (Figure 6.3).

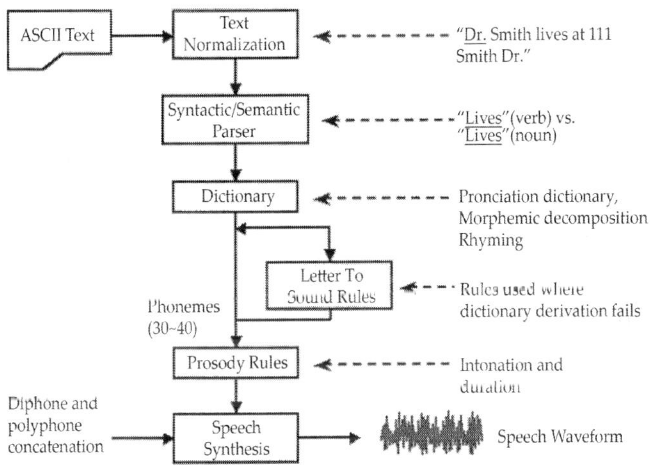

Figure 6.3. Block diagram of an English text-to-speech system.

Souce:https://www.sciencedirect.com/topics/computer-science/speech-synthesis.

Aside from the propensity for comparable voices to be preferred, early research suggested that real human speech is superior to and more comprehensible than computer-generated speech. Current findings, although, have cast doubt on such results as the effectiveness of speech synthesis increases. Listeners don't prefer genuine speech over synthetic speech when the origin of speech is a computer instead of a human, according to two trials conducted by Stern et al. (Stern et al., 2012). Indeed, the increasing humanness of smart speaker voices appears to promote excessive expectations of such machines' emotional and intellectual powers, contrasting their limits (Zhang et al., 2021).

A few of the largest outstanding issues of text-to-voice assessment, according to speech synthesis experts, is indeed a lack of qualitative understanding of consumers' demands for various application scenarios. Cambre et al. developed a study paradigm inside the Computer-Supported

Collaborative Work community, suggesting that speech design needs for smart devices have been affected by and vary between devices, users, and situations (Cambre et al., 2012). The majority of speech evaluation research to date has focused on the general usage cases of voice assistants and smart speakers, with the most widely employed metrics being linked to listening experience. For instance, comprehensibility, intelligibility, and other subjective metrics like favorability are used to evaluate voices for long-form texts (e.g., audio novels) (Cowan et al., 2017).

Whereas the preceding research focuses on how the consumer perceives voices whenever listening to speech material, consumer preferences for voices who provide their material, as well as the social consequences of those choices, have gotten far less attention. On the subject of accessibility, studies on augmentative and alternative communication (AAC) devices found that synthetic voices are difficult to use for self-expression because they lack expressivity. Enabling users' conversational tempo, identity presentation, and personality expression, which restrict their capacity to honestly articulate themselves, is one of the most difficult difficulties (DePaulo, 1992). Improvements in speech synthesis have started to address technical challenges linked to expressivity in current years. with several professional text-to-speech engines now creating more human and emotion-rich voices (e.g., IBM Watson, MaryTTS). Fiannaca et al. created two interfaces for adjusting expressive speech characteristics like picking and refining voice moods and concentrating on AAC output. HCI academics have also investigated several methods for rendering emoticons with a speech to improve the social experience of voice commands (Fiannaca et al., 1995). Ultimately, the availability of voice font technology, which uses machine learning to build a synthetic speech from a recording of the user's voice, allows for a greater variety of voice alternatives. Such developments, taken together, pave the way for speech to join mainstream social-interaction applications like social networking. However, there is a scarcity of studies on synthetic voice selections for social media. Recognizing this gap, we want to look at the criteria for displaying personalized content on social media using voice design (Doyle et al., 2019).

6.3. SOCIOPHONETIC CONSIDERATIONS OF VOICE

Whereas voice design for social media material has received little attention, it is well recognized that voice has an impact on social interactions in real life. The area of sociophonetics, which combines phonetics and sociolinguistics, investigates this phenomenon, particularly how social data has been received through phonetic features. Social categories and personality qualities may be derived from aural information in a pretty consistent manner, in addition to the spoken material (Drager, 2010). People may infer a speaker's socioeconomic status, age, sexuality, and gender, and ethnicity, only by listening to them talk. Several researchers have found that the tempo, amplitude, and accent of a speaker's speech affect perceptions of their personality qualities. Many suggest that people's perceptions of social data, particularly gender, are more akin to a spectrum than to strict classifications. Listeners' perceptions about the speaker, on either hand, alter how they receive social data from speech (Casares Jr, 2022).

Sociophonetic findings have currently been included in speech technology research in the Human-Computer Interaction (HCI) literature. Sutton et al. for instance, presented three sociophonetic-based design techniques for more accessible and natural speech user interfaces: individuality, context awareness, and diversity (Sutton et al., 2001). Such methods have focused on how speech must be tailored for information consumption. Instead, our research looks at these issues in terms of how people want their content to sound. Because synthetic voices are so adaptable, social media users may hide or expose elements of their identity that have been generally associated with their genuine voices. One thing to look at is how individuals think about using this choice to portray themselves online (Elias et al., 2017).

6.4. ONLINE PRESENTATION OF SELF

CSCW, HCI, and the sociology of science and technology have long explored how individuals let others view them on social media. Establishing social connections, disclosing identities, and changing statuses are all common social media behaviors. Previous research has looked into how social media users make decisions about their identity, audience management, privacy control, and content management with Hogan's exhibition technique and Goffman's theatrical metaphor being used (Ellison et al., 2006) (Figure 6.4).

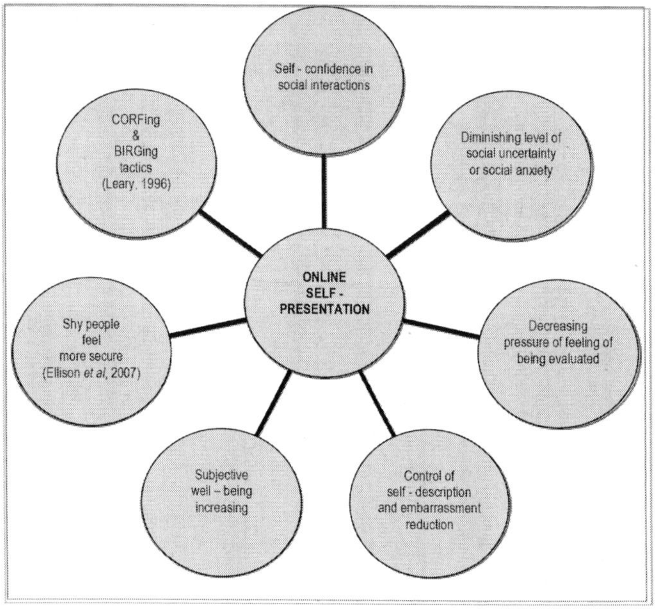

Figure 6.4. Online self-psychological presentation's correlations.

Source: https://www.researchgate.net/figure/Psychological-correlates-of-online-self-presentation_Fig1_265982045.

Dramaturgical method of Goffman employs performance and stage as metaphors to show how a person's self-presentation is chosen across situations. The term "front stage" refers to situations in which a performance is delivered in front of an audience, as contrasted to the private "backstage." People possess audiences on social media, just as they do in real life. Social media interactions, on either hand, "compress many audiences into one setting" and present users' material in a way comparable to an exhibition, unlike physical encounters that normally have specialized audiences (Fiannaca et al., 2018). To deal with presenting issues in such "exhibitions," users often employ crowd control approaches. Many individuals, for instance, create numerous accounts, only publish items that are non-offensive to the biggest audience (referred to as the lowest common denominator effect), and carefully conceal data from diverse audience groups. Using these strategies might make it more difficult to strike a balance between personal authenticity and audience assumptions. Mainstream social media frequently promotes +ve self-presentation, which can sometimes foster unfavorable social comparison between consumers and limit an individual's ability to freely represent themselves. Consumers may show themselves more truthfully and disregard social constraints from the social comparison which is common on social media via utilizing false accounts, or "finstas," with only close friends, according to Xiao et al. (Xiao et al., 1995). People generate social media material not just for others, but as well as for themselves, according to Zhao et al. to construct an "archive" of the significant moments of their lives (Zhao et al., 2000).

Nonverbal cues are frequently employed in dramaturgical analysis for online situations, although online self-presentation has mostly focused on how users choose textual or graphical information. When picking synthetic sounds to express oneself, little is recognized as to whether people use impression control tactics. Is it possible that the accentuated difficulty in handling impression management via collapsing contexts presents unique issues for voice design and creates a place for voice-customization exploration? If that's the case, how do consumers recommend adapting their synthetic voice presentations to address these issues? In our research, we look into these issues (Goffman, 1978).

6.5. VOICE-BASED SOCIAL MEDIA

Various social media platforms are beginning to accommodate the distribution of voice-based content. Users may also listen to and record voice messages on certain recording-based voice boards. Gurgan Idol, Baang, and Sangeet Swara, are three research-based voice forums aimed at people with poor literacy and socioeconomic hurdles to internet knowledge. Such services have been accessed through toll-free phone calls in regional languages, and the study has emphasized financial sustainability, usability, equality, and integrity. Shoot words, Clubhouse, Hear Me Out, and Adult are examples of recording-based auditory networking networks that cater to a wider audience (Greene et al., 1986). These websites claim that incorporating audio contact into online interactions may make them more genuine, engaging, and easy. Moreover, the fact that such services only offer recording-based interactions might be an issue for consumers who are unable to record their voice, like to write material utilizing text-based approaches, or wish to generate content that may be consumed in many modes. Another source of worry is the privacy implications of voice recordings. As a result, the emphasis of our research is on how contemporary social media consumers might represent themselves online using a synthetic voice (Govind and Prasanna, 2013) (Figure 6.5).

Figure 6.5. Voice-based social networks.

Source: https://www.scoopearth.com/voice-based-social-networks/.

6.6. METHOD

We did semi-structured interview research with fifteen respondents to see how they respond to the notion of employing personalized synthetic voices to deliver their material on social media. We used audio instances to urge respondents to imagine a variety of synthetic speech features, some of which contained the participant's social media information. Participants were asked about their reactions to the notion of utilizing a profile voice in general, the voice features offered, and their preferences for personalized or default voices in various situations. Our university's Institutional Review Board gave its approval to this study (Alonso Martin et al., 2020).

We utilize the terms bio information and profile interchangeably to refer to the opening explanations that social media consumers submit about themselves on their profile pages and profile to refer to the overall word that encompasses both posts and bio information (Halderman et al., 2004).

6.6.1. Participants

The study included 15 active social media consumers who had certain experiences with voice interface technologies. We used four social media channels (Twitter, Reddit, Instagram, and Facebook) as well as word of mouth to recruit for diversification. Respondents had been chosen depending upon the following parameters (Zhang et al., 2021):

- They should have at least two postings on Instagram, Facebook, Twitter, and LinkedIn as well as one written paragraph of bio-data;
- Minimum once a week, they should utilize social media sites;
- They should have blogged a minimum of once in the previous month;
- They should be familiar with smart speakers or voice-activated systems like Google Home, Apple Siri, and Amazon Echo (Hu et al., 2019).

The participants were between the ages of nineteen and forty-seven. Table 6.1 shows such information, as well as self-reported gender, race, and major social media site. All of the participants were volunteers who had been given a $20 gift card in exchange for their time (Hogan, 2010).

Table 6.1. Participants' Demographics and First Voice Choices Amongst Five Voices Differed Only in Reported Age and Gender. Ethnicity, Age, and Gender are Stated in the Participants' Terms (e.g., "Woman," "Female")

ID	Gender	Age	Ethnicity	Primary Social Media Platform	Preferred Voice
1	Female	20	African American, Caucasian	Instagram	Older masculine
2	Female	20-something	Black, white	Twitter	Gender neutral
3	Female	30	White	Facebook	Younger feminine
4	Genderqueer	20	White	Instagram	Younger masculine
5	Female	19	Black	Instagram	Younger feminine
6	Male	25	Mixed race, but mostly white	Instagram	Gender neutral
7	Female	47	African American	Facebook	Older feminine
8	Male	20	African American	Twitter	Younger masculine
9	Female	20	Asian	Twitter	Younger feminine
10	Woman	20	Caucasian	Facebook	Younger feminine
11	Male	23	Half black, half Filipino	Instagram	Younger masculine
12	Female	21	Asian	Instagram	Younger feminine
13	Female	20	Asian, Caucasian	Instagram	Younger feminine
14	Female	21	White	Instagram	Older masculine
15	Male	32	African American	Instagram	Older feminine

6.6.2. Procedure

The interviews had been scheduled to last up to sixty minutes and had been done using video conferencing software from a distance. Participants provided their bio-data as well as two examples of posts from their basic social media accounts before the event. All interviews were performed by the same member of the research team for consistency's sake. Every interview consists of four sections (Hunt et al., 2018):

Participants had been 1st questioned about their demographics, language skills, social media and technology usage, and voice technology experience (James et al., 2018).

Secondly, we generated emotions to the concept of interacting using social media material by a speech by presenting the scenario below: " Assume that other people, in addition to yourself, may utilize voice-based technologies such as smart speakers to listen and interact with social media. Think about how you want your articles and bio-data to come across, as well as what other people might think if they heard them (Kane et al., 2017). To provide respondents with concrete examples and assist them in imagining a variety of different speech alternatives, we played a section of their own social media profile data using five different sample voices. Old and young female and male voices, as well as a vocally androgynous (gender-neutral) voice, had been chosen to reflect a variety of perceived sexes and ages. Natural Reader and IBM Waston had been used to create these 5 voices. All audio recordings were kept to a length of around ten seconds, that's plenty of time to read normal update-type postings (e.g., 240 characters on Twitter). The site Unicode Common Locale Data Repository was used to transform all expressions into spoken words (Joinson, 2008).

Participants responded to their profiles read aloud by the 5 voice specimens (arranged randomly for each participant) without being told about the planned gender and age disparities. Participants had been asked to characterize the voice, score their degree of agreement with the statement "I will prefer to utilize this voice for my social media profile" on a five-point Likert scale, and write a justification for their assessment after every clip (In Section 4, we concentrate on the qualitative explanation, while the rating data is presented in the supplemental materials). We also ask respondents what elements they assumed while assessing the voice specimen and which aspects of their identities (race, gender, etc.) they will be comfortable having a synthetic voice represent (Koradia et al., 2013).

The 3rd section of the study concentrated on more sophisticated synthetic voice features. We tested reactions to potential advanced traits by playing two sets of sample voices with distinct accents (British English and Indian English) and emotions (furious, pleased, sad, and afraid), all created with Voicery. "The swift brown fox hopped over the slow dog," the accented specimens stated, while the emotional samples said, with the phrases shifting to fit the emotion (Lee et al., 2007). "It was not simply a (good or bad) dream." It caused me (depressed or angry or frightened or happy). We stated that the purpose of such samples was to inspire participants to consider advanced voice characteristics and that synthesizing of such aspects will likely increase in the following years. We questioned participants for their initial impressions after reviewing the two sets of clips, as well as whether they thought being capable to modify emotions and accents will be beneficial or not for their social media voice accounts. The interviewees had been questioned more broadly about what qualities of synthetic speech if there are any, may be beneficial for social media consumers to tailor the sound of their accounts, as well as what voice they will choose for their profile and why (Krämer and Winter, 2008).

Finally, the interview concluded with questions on the differences between default voices (such as Alexa or Siri) and customized voices for social media accounts, as well as if voice choices will alter in various situations (e.g., for various platforms or kinds of content) (Lee et al., 2003).

6.6.3. Data Analysis

Based on our interviews, we conducted the thematic analysis with the goal of better understanding voice features that are significant to participants' consciousness and how circumstance impacts profile voice preferences. To gain a general idea of the subject matter, the first author of the study read all of the transcripts to get a comprehensive picture of it, and then he or she compiled a list of preliminary codes that included both inductive and deductive codes (Luger and Sellen, 2016). After that, the first two writers each independently coded all of the transcripts, and then they would meet every two or three participants to discuss the coded transcripts and memoranda, resolve any issues that arose, and iterate on the structure of the codebook. We found five major topics after the 1st round of coding: criticism on emotional voices, comments on accented voices, general answers, variables to consider when assessing a voice, and how and when to customize profile voices (codebook that was made available for use as supplemental material). The 1st two writers went over the codebook again, reviewed it, and regrouped it, which resulted in the final ideas we provide in this work (Bakken et al., 2017).

6.7. FINDINGS

We explore voice issues specific to self-presentation on social media here (Marwick and Boyd, 2011).

6.7.1. Voice Choice by Content Producers

The majority of previous talks about synthetic voice choice centered on listeners' experiences. Despite this, individuals produce a large portion of web material and practically all of the social media. What voices must users employ to communicate their material (James et al., 2020)?

When prompted how they would rate synthetic voice alternatives for their own social media posts, respondents stated several aspects that essentially consist of two types: how the voice portrays personal qualities and the general effectiveness of the voice. Table 6.2 covers all of the criteria, comprising those that were specifically specified in interview questions (gender, age, mood, and accent) and others that surfaced suddenly throughout the interviews (Matias, 2020).

Table 6.2. Voice Personalization Preference is Influenced by Several Factors

Category	Factor
Representation	Personality Gender Pitch Age Accent Emotion (of content)
Overall Quality	Naturalness Accuracy and Clarity Timbre
Other	Basic voice characteristics (speed, cadence, volume) Non-text sounds (e.g., music, animal sounds)

Several reasons came up, but people weighed them in various ways. The capability to organically convey who they are and appropriately represent the content of their statements was maybe most important. We explain why and how participants felt specific characteristics of synthetic voices had been significant to the presentation of their information in the sections below (Mayer et al., 2003).

I'm looking for a representative voice. The need for a representative synthetic voice was mentioned by all interviewees. "I'm reading such things (its own social media postings) in my thoughts like I'm listening to my voice," P4 explained. If there's somebody there, it'll be odd." Participants emphasized constant personal features (e.g., personality, gender) as well as the sentiment of their material when discussing how voices will or will not reflect them (e.g., a sad or happy post). Table 6.1 shows that, with a few exceptions, the favored voice's age and sex seemed to match the participant's age and sex (McGregor and Tang, 2017).

Most such interviewees remarked on whether or not the apparent personality of a voice fit their social media identity ($N = 14$) when analyzing the voice recordings. The synthetic voice wasn't really "quirky" (P3), "positive" (P1), "friendly" (P15) enough, or "confident" (P5), for instance, indicating a personality gap. Certain respondents ($N = 5$) valued accurately portraying their personalities, as a single respondent put it: "I don't need it to sound like someone I am not, since I am not trying to trick anyone, or have a different character" (P11) (Eitel-Porter, 2021). Others ($N = 6$) were more focused on matching the voice's characteristics to their online profile, and not as an accurate portrayal of themselves. P12, for instance, stated: "Since, as I already stated, (my post) is favorable." Although my personality has highs and lows, and I only express the positive side of things, which isn't my personality; it's simply a component of it." On average, individuals prefer to project a favorable self-image on social media. "All of my words or postings are quite, I will say, lively or upbeat," said one person. 'Oh, I completed my skydive!' or 'Oh, I did my internship, so great!' yet he seems so exhausted, that makes me wonder if he's mocking my job?" (P12) (Ilag, 2020).

Gender and Pitch: Almost all interviewees desired their profile voice to sound such as the sex they identify with, as seen in Table 6.1, albeit to varying degrees. "Even though this voice has the highest quality of any other voice that I have heard so far, it is still a female. Furthermore, how the female is delivering the Figurative language that she is speaking is superior; however, I will still want it to reflect who I am when it comes to my profile," one participant said. I am a man, and I do not speak like that." (P15) (Mirenda et al., 1989).

People, on either hand, have diverse views on how significant gender is to their identity. P1 remarked that two respondents had been ready to prefer voice quality over gender-matching the voice:

"I believe (gender) had a significant role in how I felt it associated with me." But after hearing the last voice, I didn't seem to care as much about gender. So I stopped paying attention to gender because I just preferred that voice. (P1) (Mori et al., 2012).

One characteristic of voice that is related to gender is pitch. About 50% of our respondents expressed a desire for a pitch that reflected their gender identity. Nonetheless, inside a gender group, a pitch might differ. Whenever the voice not only reflected their gender identification, and also expressed that they were in comparison to the typical voice in that group, our respondents were pleased. "Although I am a girl with a deeper voice than is typical for my gender, it did sound deeper than number 4, which was the younger female voice," P2 explained. So it's something I enjoyed because it's something I can relate to. "A few folks were dissatisfied with how feminine or masculine the gender voices sounded in this style. P6, for instance, when asked to explain his ideal voice, expressed a need for more gender-neutral options: "something is masculine, but not every male I think." This shows that a broad range of pitches might be more beneficial than a limited selection (Mullennix et al., 1995).

Age: Whereas not as relevant as gender or personality, our respondents' ages were regarded to be an element of their identities. Respondents found it difficult to maintain their online character whenever a voice sounded very young, as in: "I suppose I don't sound like an elderly woman." Yeah, as I have stated, it's just not compatible with how I perceive my voice to sound" (P4). Most respondents, on the other hand, were unconcerned with tiny age differences: I believe it comes through when you attempt to adapt it to your personality or how you portray yourself to be, although I didn't use my age as a deciding factor" (P9). Because our participants were mostly young individuals, the results may be different for older consumers (Munson and Babel, 2007).

Accent: Respondents who regard accent to be a lifestyle believed that including accents in their speech presentations will help them express themselves better, as P10 put it:

"Well, I'd feel more relaxed and secure having the way I talk portrayed, mainstream media, and contain a voice that truly fits the manner I speak, instead of one that matches the approach everyone else speaks." (P10) (Zhang et al., 2021).

We tested interviewees' reactions to synthetic speech sentiments by playing four voice specimens that sounded angry, sad, pleased, and afraid, in addition to the preceding identity-related criteria. The ability to adjust voice sentiment to fit certain social media postings, according to all participants, might aid in the delivery of meaning. "I believe sentiments are what's lacking in text and social media," P13 explained, "and being able to transmit that in a more audio manner may provide an added dimension to what people have been missing in their social media." "I want everyone to hear how I perceive things," P11 said with passion. That, I believe, significantly alters the game" (Niedzielski, 1999).

Respondents also frequently raised their hands when the voice's sentiment or tone didn't match the substance of the article. After hearing a voice that he interpreted as unemotional, one respondent particularly commented how disturbing the discrepancy between content and voice might be: "They may see in the post which I had a nice time, however when you hear it that manner, I feel it detracts from the overall impression and draws attention to that voice." (P15). "If it had been an exciting time, I will like my voice to be more energetic and perhaps put extra emphasis on specific things," P15 remarked when questioned about voice options for Instagram. I'd like it to be neutral if it's sad, but it depends on the image" (Farrell, 2014).

Despite the significance of emotion personalization, there had been still reservations about using algorithms to simulate and display emotion: "It's highly subjective, emotive." When individuals post truly terrible things, it's possible that certain individuals may find it amusing or I don't want the voice assistant to decide on whether it is a sad or joyful post. This is one murky area about which I'm unsure." (P12) (Pearl, 2016).

Several respondents proposed other sentiments that might be good to depict in addition to the ones we offered. Four people suggested sarcasm, for instance: "I believe that will be incredibly useful, particularly if you might get over the sarcastic tone." That will save a lot of people a lot of difficulties since sarcasm doesn't translate well in messages" (P2) (Brunet, 2021).

Wanting a voice of outstanding quality. Voice quality is crucial to the listening experience and content comprehension and has therefore always been an essential measure for determining speech synthesis. We noted additional voice-quality related considerations once interviewees analyzed voices for their social media content in addition to listening pleasure and comprehension. The most important point is that a voice that is unnatural or indistinct cannot sound representative despite the manipulation of other qualities. When asked about their concerns regarding voice quality, respondents most frequently highlighted accuracy, clarity, and naturalness (Doraiswamy et al., 2019).

Naturalness: The degree to which a synthetic voice must sound human differs by application. Almost all respondents ($N = 13$) favored a voice that sounded natural; it was a trait that we did not directly inquire about, highlighting the relevance of genuineness for voice-based social media distribution. The primary participants all agreed that naturalness was the most important quality of a good voice, even to the point of sometimes surpassing generalizability. For example, participant P8 stated, "I believe that out of all the voices, the younger male voice was the one that sounded the most humanlike (Pullin and Hennig, 2015). It ticked the most boxes that were essential to me, and the ones that it didn't check were less significant because it was the one that checked the most boxes." I believe that to be the most significant." Respondents disliked voices that seemed too flat for fear that the voice will not effectively express the meaning of the text and will hinder the user's capability to show oneself, like: "Will does not sound right. The voice that reads my bio and my article sounds monotone, and I believe it distracts from whatever I want people to need focus on it. The voice is still rather artificial, and it does not portray who I am as a person (P15). Respondents characterized a natural-sounding voice as having suitable intonations, a natural movement of words, and a genuine tone (e.g., P11: "I loved how it had diverse tones and inflections") (Porcheron et al., 2017).

Accuracy and clarity: greater than 50% of our participants selected voices depending on how properly they talked ($N = 8$). For instance, one respondent was concerned that the speaker's speech was too rapid for listeners to comprehend, which is related to clarity: "It was rapid. If I had been attempting to comprehend something stated to me, I might prefer a slower rate of speech (P2) (Scott et al., 2019). In addition to the intelligibility of the speech, respondents were particularly concerned about the correct pronunciation of terms. For instance, four individuals had been concerned about the mispronunciation of names or concepts on their profiles: "It stated, 'My dog's nickname is Milan,' but it came out as 'Mulan' or whatever. I observed that some terms were being mispronounced. I suppose. I don't know if it's the punctuation, but the sentence didn't flow properly (P15) (Purnell et al., 1999).

Other voice-related factors. In addition, less frequent ideas included the ability to alter cadence, voice speed, and loudness, as well as the support of non-text components like animal noises, strange vocal distortions, and distinctive background music, as seen in Table 6.2. Collectively, they represent social media users' demand for greater control over their online presentations and their wish to explain themselves creatively (Robinson, 2007).

6.7.2. Voice Choice Across Social Contexts

In the preceding section, we examined choices for synthetic social profile voices generally, although respondents also considered how these choices can vary depending on the social situation. We described four consistent factors: (1) the need to seem real and constant irrespective of social environment, (2) the desire to modify voice tones and phonetic style based on (3) content type and (4) audience, and (4) the distinctions between public and private social media postings (Sebastian and Ryan, 2018).

Firstly, respondents favored maintaining the consistency of the major qualities of their synthetic voice (*N* = 13) (Oosterom, 2021).

P5 underscored the significance of consistency by considering how her contacts and friends will respond to voices that didn't sound like her: " If I had picked younger female voice option number one or younger male voice option number three, for example, people would have been like, "What is she doing?" That doesn't sound anything like her at all." (P5). In addition, P3 commented on the potential for confusion brought about by switching voices: "And I think it will be pretty complicated for the audiences, you know, if one day Lori (pseudonym) has the voice of an elderly man and the next day Lori (pseudonym) has the voice of (inaudible), you know, I think that it's probably easiest for everyone if it's consistent" (Sutton et al., 2019). Secondly, despite the importance of keeping a voice's general consistency, several respondents (N = 13) said it will be advantageous to change certain qualities based on a particular material. For instance, P12 stated, "If we move to a new voice or sound? That is odd. However, if it's the same individual with a different tone, that's OK." Such little alterations were viewed as particularly beneficial for expressing emotion (N = 11), such as: "If it's constant with the probable exception of adding that type of emotional intonation, I believe you have the same voice" (P3). For some respondents, the ability to perceive the emotions of others led to enhanced social ties and mutual understanding. P2 believed that emotion brought individuals "closer together" and noted (Stern et al., 1999):

"I go through stages with my tweets, in which if I'm feeling extremely down for a week, all of my tweets would be incredibly depressing. The moment I regained my equilibrium, though, I will want to experience something joyful. Such that my timeline is not a perpetual source of grief for everyone." (P2) (Stern et al., 2006).

Several respondents want emotional voice synthesis only for social media posts, and not for functionalities. The majority of respondents said that content that was neutral, objective, or supplemental, like their bio information, and esthetically dominant material, like Instagram posts, didn't require customization: "primary profile information will likely be conventional as if I were delivering a lecture; not very emotive, but extremely official." (P2) (Strand, 1999).

Thirdly, people perceive themselves and engage uniquely across the various social media platforms; as a result, respondents stated considerations for the people's expectations and the culture of the community of each platform they use (Fox and Wu, 2021). For career-oriented networks like LinkedIn, around 50% of our respondents ($N = 7$) said they will employ a more professional voice. "If there were additional possibilities that popped out, perhaps one that seemed a little more official, then I might select that for a Facebook or LinkedIn sort of situation," P9 explained. For such kind of circumstances, something that's more professional or even-toned" (P9). A few people had similar sentiments about using a more professional tone on Twitter, whereas others thought they had more leeway in how they presented themselves (Tamagawa et al., 2011).

P2 wanted to employ a more optimistic voice on Facebook to communicate with family members with whom she wasn't familiar enough to talk about serious subjects (N = 6) (Thomas, 2002).

Ultimately, the personal features to reveal through a synthesized voice depend on how publicly or privately the content of the post will be for specific people. P13, for instance, had a personal account in which they might imagine utilizing a personalized voice, but "I wouldn't feel as happy letting somebody hear my voice if they weren't a buddy, a public account." In that case, I suppose I'd use Alexa's (default) voice." P1 also stated that she will feel more at ease utilizing "sad or furious" voices on a more personal Instagram account, but that "my major [account] which I use more always shows, such as, the cheerful highlight." (P1). P6 also mentioned that some sites, like Reddit, are intrinsically more anonymous, and that it will make more sense to adopt a preset voice there since ."I will not necessarily need everyone to understand further about me" (Tusing and Dillard, 2000).

6.7.3. Challenges of Voice Customization for Self-Presentations

Whereas virtually all respondents expressed enthusiasm about accessing social media with such a new method of engagement, there were some reservations regarding identity-based voice personalization. Three major problems are discussed below: misrepresentation, risk of stereotyping, and the influence on usability. Stereotyping may occur due to customization: Greater than 50% of our respondents (N = 9) highlighted the likelihood of stereotyping, especially about accents and genders (Vashistha et al., 2019). According to P4, "I might see individuals utilizing it intentionally as well, such as make fun of accents," and P9, "Particularly with English, such as Asian accents in English or whatever," an accented voice can be incorrectly linked with these groups of people. Some might argue it's overly stereotypical, or that synthesizer sounds strange." Several respondents had been particularly concerned that this chance will be exploited for cyber-bullying. "This will give a chance for the world's mean-hearted individuals to mock and tear others down, including all who have speech impairments or disorders," P5 added (Tigwell et al., 2020).

Several interviewees mentioned that conveying ethnicity via voice might be difficult, which is connected to accents. Two white respondents (P4, P14) asked what a "white-sounding" voice may sound like in answer to a question as to which elements of their identity they will feel easy reflecting with a bespoke voice, whereas one Asian respondent shared a similar concern. Two persons of color (P7 and P8) expressed concern about maintaining stereotypes by using voices that matched their ethnicity: "It will make me uncomfortable if a voice sounded extremely African American in regards of being too stereotypical since it will make me feel like they were overdoing it" (P8). P15, who was recognized as African-American, believed that managing a broad range of voices might help to dispel stereotypes, saying, "Perhaps if we all had voices that made it sound such as mine or sounded such as the various versions of people will not have that set stereotype of what an African-American man will indeed sound like" (Kim et al., 2020).

P13, for example, expressed similar difficulties with gender stereotypes: "I feel bad presuming that such voices sound like a particular gender." P4, who self-identified as gender-queer, expressed reservations about using a synthesized voice to convey gender. Such results imply that social stereotypes have a strong influence on speech recognition, necessitating a greater understanding of this phenomenon in voice user interface design (Vasquez and Lewis, 2019).

There is a risk of deception when customized products are used: A small number of responders (N = 3) were concerned about the possibility that personalized voices could inappropriately emphasize the seriousness of their remarks and accidentally upset listeners: "I don't worry about being provocative," one of the responders said, "but usually speaking, I try to post stuff that I understand won't irritate everyone since the problem is, I do have relatives on there, and I won't create Fights online for no reason" (P2) (Pickup and Thomson, 2009). To prevent this circumstance, several respondents will strive to employ a neutral or preset voice: "So then it isn't like no one may evaluate what voice I used, since it's just so neutral that it doesn't matter, such as I nearly did not choose anything" (P14). Two respondents also expressed their reluctance to use a voice that sought to be real but failed to achieve this goal, leading to a quality known as the "uncanny valley" that might affect how the information was perceived. In this particular scenario, the use of the default or "normal" (P12) voice had been recommended (Cambre et al., 2020).

Potential usability implications: Significant usability problems occurred with the potential ramifications of individualized voices, that would require further investigation in future projects. Four individuals remarked on the hardship of personalization as either of those. P14, for instance, believed that unless she was "very interested in making my profile more entertaining or like much more tailored to me," she would "certainly want to use a standard one" if she did not have time to put it up (Garner et al., 2010). P9 had a similar attitude, admitting, "I guess I will certainly only do it for perhaps a few blogs for me individually." Whenever it comes to media, I'm a bit of a slacker, but I believe it will be beneficial in the long term." "Excessively personalization will make it difficult for (a screen reader client) to understand the material, which is far more essential than the voice where it is conveyed," According to P3, another (observed) respondent had been concerned about how individualized voices may affect usefulness for blind consumers depending on screen reader audio (Zhang et al., 2021).

6.8. DISCUSSION

We investigated how social media consumers might want to offer their material on voice-based platforms via interviews involving fifteen people. Whereas previous research has looked at how people like to interact with and consume voice-based information (e.g., various forms of texts), this is the 1st study to look at how people choose to offer their own socially driven content. Our results suggested that such tailored voices must complement, or at minimum not interfere with, social media users' online personalities and the precise material being delivered, in addition to conventional voice characteristics like likeability and intelligibility. Users prefer a standard, basic voice to depict them whenever a customized voice fails to meet such criteria, and whenever voice personalization interferes with other goals like usability. Such findings contribute to current debates on voice interface design for social software by providing empirical knowledge about what renders a voice appropriate to portray a social profile (León-Montaño et al., 2020).

In the parts that follow, we'll go through how the results from this research contribute to the theoretical understanding of voice interaction design and online self-presentation, as well as how and when they might extend to different sorts of voice apps. We finish with the design requirements for a better online synthetic speech presentation (Waterloo et al., 2018).

6.8.1. Online Self-Presentation and Synthesized Voice

The participants in our study preferred to evaluate their synthesized voice presentations having a collection of imagined audiences in mind, adopting tendencies of online management of impression. Whereas previous research has looked at self-presentation on social media using textual and visual information, this study builds on previous work by identifying commonalities and revealing distinctive features of voice-based impression management. Accent, age, and gender are just a few of the components of identification that may be expressed rapidly and clearly via voice (Zeng et al., 2017). Synthesized voices enable consumers to be more attentive to their presentation preferences, presenting them with unique and explicit chances for impression management on social media, since these characteristics can't be as simply retrieved via textual material. For instance, to preserve a constant impression and prevent being viewed as inauthentic, our respondents reported specimen sounds depending upon how much the voice sounded like them as well as how various people will interpret that voice. The key to maintaining an authentic voice delivery, according to them, was to transmit the proper personality and crucial components of their identity. In addition, we found how the lowest denominator impact could play out in synthetic voice choice: whenever in question about the acceptability of a voice, several participants chose what they perceived to be neutral defaults (e.g., Alexa's or Siri's default voice) (Sijia Xiao et al., 2020).

Other frequent impression management techniques were also evident in the individuals' voice selections. Self-enhancement (e.g., the endeavor to show oneself in a positive light to others) and conformance to the norm in such a given social environment (e.g., a specific social media platform) are two instances of such behaviors. For instance, it is common knowledge that social media platforms encourage users to present a favorable image of themselves, and the voice preferences of our participants supported this cultural norm. Optimistic and p Positive voices were frequently chosen by the participants, even though these voices didn't adequately represent the respondents' emotional states (Zhao et al., 2013).

Participants contemplated how they may adjust synthesized voices to adequately represent their identity on a certain platform and to enhance the tone to make it fit for a particular post. This is analogous to the way consumers of social media adopt various verbal and visual styles throughout situations (Susini et al., 2004).

However, despite advancements in the synthesis of speech, there are still some limits that make using a synthesized voice come with its own set of difficulties. Participants underlined the importance of high-quality synthesized speech and expressed concerns on how constraints like inaccurate pronunciation, inadequately precise expression of emotion, and the "uncanny valley" impact might affect listeners' impressions of a user's information and identity. The need for higher-quality synthesized speech was also stressed by participants (Smith, 1996). Incorrect pronunciation can make a person come out as less professional and, if it involves the mispronunciation of the name of a relative or friend, it can also generate friction in relationships that are already well-established. A poor modification of an expression of emotion or accent has the potential to not only distort the understanding of the message but also to insult other people. As the abilities of speech synthesis continue to advance, this would be necessary to review such problems (Lindau et al., 2014).

Currently, probing questions for additional work to be done in the theory of voice design, especially in certain situations, and the debate that was just presented may guide the expansion of the present theory to consumers as developers of a voice. Cambre et al. provide a methodology for developing voices for intelligent devices by taking into account the perspectives of the device, the context, and the consumer (Cambre et al., 2001). Even though our emphasis on social media information is typically not reliant on the device being used to access it, and it is still important to consider the context and the consumer. Regarding the user lens, Cambre et al theory's views computers to be social actors and advises that voice designers consider the impact of users' particular choices for the kind of social contact that is being modeled by the computer (e.g., complementing or matching consumer properties). The question that is now being prioritized is how the user may make the personality and other social aspects of the computer or agent appealing to them. Whenever the design issue changes to how to help consumers express themselves as well as their social material, a voice designer must evaluate how (for instance, via gender, personality) and to what degree the user wants to expose their personas and identities. This can be done by asking questions like "how much do you want to reveal about yourself" (Dubiel et al., 2020)?

Cambre et al. also consider the larger historical and cultural context in which the usage of smart devices occurs while conducting their analysis of voice design (Cambre et al., 2001). Similarly, our research has revealed that people's preferences regarding the tone of their voices change depending upon the social context, which may involve a variety of groups, platforms, or particular posts.

Wiki pages and other platforms that allow for anonymous involvement, such as Reddit, are good examples of the type of platforms that support user-generated information which is not intimately correlated to an individual's identity. We hypothesize that consumers of these types of platforms would then likely have a unique set of voice prerequisites (Valkenburg and Peter, 2011).

6.8.2. Design Considerations

Our study takes a prospective approach to locate potentially fruitful avenues for further research and design investigation in the future. Our research demonstrates that a rise in voice-based interactions may provide consumers of social media platforms with additional control over their self-presentation if changeable voices were used for spoken personal details. This would be a very welcome development. However, there are also significant usability, technical, and ethical difficulties that would need to be resolved. In this article, we discuss certain design issues that should be taken into account for upcoming work on social context, and self-customized speech synthesis systems (Kobayashi et al., 2010).

Options for customizing the voice. What alternatives, if any, must designers give between all conceivable synthesized voice qualities, and how? The general quality and genuineness of the voice, according to our research, is a necessary base for any customization. Furthermore, due to existing limits in the synthesis of voice, perceived quality can act as a roadblock to the acceptance of fine-grained management of expressive accents and tones shortly. New design characteristics, like the ability to add phonetic gloss to titles and other terms that have been likely to be mispronounced, might solve a few of the present technical restrictions with voice synthesis (Meimaridou et al., 2020). The difficulty then becomes how to enable consumers to modify the voice once a basic quality level has been achieved. Instead of imposing fixed classifications like age and gender, certain voice qualities must be supported along with a range (Even though certain respondents did not consider age to be as essential as other factors). If diverse accents are supported, as several text-to-speech algorithms currently do in a limited way, a complex set of choices would almost certainly be required to oppose and avoid stereotyping, as stated below. Ultimately, maybe most significantly, every voice customization ought to be optional, including designers offering a default voice and allowing users to override the system if it ever adjusts the voice to the material or user (Zhang et al., 2022).

User effort is reduced. Even though we have claimed that consumers ought to have nuanced control and fine-grained as to how a voice sounds, customizing a voice requires effort. Participants wanted a representative voice, but not needed one that sounded exactly like their own. An "uncanny valley" impact could be created by a voice that strives to sound exactly like such a real person but fails. The value-effort trade-off between providing consumers with fine-grained flexibility and having a restricted selection of predetermined voices is an interesting subject for future research. Another way to reduce setup time is to allow users to choose a small number of descriptors (e.g., 3–5) that define whatever they want to communicate about themself, and then have the system create a preliminary profile voice based on those facts (Byeon et al., 2022).

In addition to a general voice for the profile as a whole, it may be especially time-consuming to modify the phonetic qualities of specific posts or other information. When consumers are composing their material, we recommend that they have access to voice tone options that can be activated with a single click, in addition to a neutral voice or default option that is always accessible. This recommendation is based on AAC designs like the Editor of Voice setting and the Expressive Keyboard (Apple et al., 1979).

Automatic adaptation can also be desirable because it might assist in determining the suitable voice tone for usage; nevertheless, any prediction mistakes might lead to confusion and misconceptions, which is a problem that was expressed by our respondents (Graps, 1995).

Personalizing artificial voices could cause problems. In our research, significant ethical concerns about voice customization appeared. Participants, particularly non-Caucasian participants, voiced fears that voices incorrectly linked with their cultural and ethnic backgrounds will spread damaging perceptions about their identity. One of the panelists had been especially concerned about the voice choices available to those having speech impairments (Västfjäll et al., 2003). Designers should be aware of any unintentional preconceptions that may arise as a result of the development of voice technology, and they should be open about any algorithmic choices taken whenever generating voices based on users' identification labels. Users must be capable to adjust the algorithm-determined vocal characteristics when they need to. Perhaps one of our participants suggested that we shouldn't depend purely on algorithms to determine how an article or a person must sound (Akhtar and Falk, 2017).

Additionally, highly tailored, natural-sounding voices may be used to identify individuals. If users' voices have been accessible to anyone on social media, there has been a possibility that people will use them for negative objectives like deceit, cyberbullying, or criminal activity. Whereas the voice synthesis community is aware of the potential ethical difficulties associated with voice cloning [Bradbury & y Gasset, 2014], it is uncertain how personal synthesized voices would be governed (Poepel, 2008). We recommend that designers make the code of ethics and legal implications of employing speech synthesis as obvious as possible. Efficient measures to avoid damage must also be considered by policymakers. One potentially contentious policy option is to limit the usage of voice features that don't pertain to users' identities, like requiring consumers to register their unique voices before usage. The application of this law, however, will come at the expense of users' freedom of speech and expression (Bradbury & y Gasset, 2014).

In the direction of more audio material. Although our research is dedicated to delivering text data through voice, social media content frequently includes non-text elements such as emoticons, videos, and photos. Reading emojis loudly and playing sound clips of laughing or weeping are two examples of existing experiments to speak out emojis (Tanaka, 2006). When it comes to images, present solutions depend upon the alternate text, which is frequently absent. Consumers are more inclined to type out image captions if the voice-based availability of social media is used as a supplement to more traditional access, therefore boosting the availability of such services. Furthermore, the audio paradigm brings up novel avenues for self-expression, like providing trademark background music, as several respondents had envisioned. Future research would focus on how and if to support this information, and more work would be required to find and iterate on non-text aspects that could be conveyed by speech (Beilharz and Vande Moere, 2008).

6.8.3. Limitations

Our work is limited in numerous ways. Firstly, it is exploratory research in an uncharted field of study. Although we presented participants with specific audio samples utilizing their content on social media, they were still required to guess how they would utilize a future technology. Reactions from participants might be influenced by novel impact and demand factors, which might have resulted in falsely supportive comments about the overall concept of voice customization; actual adoption trends would undoubtedly vary. Reviewing the outcomes using a fully functional system is a crucial next step for this project (Repp, 1997).

Secondly, we made a conscious effort to recruit a varied group of participants, and while we had been successful in terms of ethnicity and gender, practically all of the respondents were young people between 20 and 30 years, with a majority of them being women. Along with a broader variety of respondents ages might have resulted in the emergence of new themes as well as a shift in the significance of certain themes (e.g., Perhaps the reactions to using voice to convey age will have been different) (Västfjäll et al., 2002; Samartzis, 2012). Whereas our respondents were familiar with a variety of social media platforms, they mostly utilized Facebook, Instagram, and Twitter, which restricts the applicability of our research to other platforms. Customers that depend significantly on audio interfaces for accessibility, such as screen readers and AAC consumers, must be consulted in future studies regarding their thoughts on voice modification for self-presentation. Thirdly, developing expressive synthesized voices is still a work in progress, and the instances we showed respondents, especially for accents and emotions, had not been evaluated as completely natural. Even though there had been a good response to the concept of altering sentiment to a certain level, the lesser quality of voice specimen may have influenced respondent answers. Lastly, research of other kinds of user-generated material, like crowd-sourced review platforms and blogs, can be useful in determining how consumers' preferences and requirements for audio-renderings of material vary (or remain constant) across media types (Fröjd and Horner, 2009).

REFERENCES

1. Addington, D. W., (1968). *The Relationship of Selected Vocal Characteristics to Personality Perception* (3rd edn., pp. 492–503).
2. Akhtar, Z., & Falk, T. H., (2017). Audio-visual multimedia quality assessment: A comprehensive survey. *IEEE Access, 5*, 21090–21117.
3. Alonso Martin, F., Malfaz, M., Castro-González, Á., Castillo, J. C., & Salichs, M. Á., (2020). Four-features evaluation of text to speech systems for three social robots. *Electronics, 9*(2), 267–269.
4. Apple, W., Streeter, L. A., & Krauss, R. M., (1979). Effects of pitch and speech rate on personal attributions. *Journal of Personality and Social Psychology, 37*(5), 715–720.
5. Aronovitch, C. D., (1976). The voice of personality: Stereotyped judgments and their relation to voice quality and sex of speaker. *The Journal of Social Psychology, 99*(2), 207–220.
6. Bakken, J. P., Uskov, V. L., Kuppili, S. V., Uskov, A. V., Golla, N., & Rayala, N., (2017). Smart university: Software systems for students with disabilities. In: *International Conference on Smart Education and Smart E-Learning* (Vol. 1, pp. 87–128). Springer, Cham.
7. Beilharz, K., & Vande, M. A., (2008). Sonic drapery as a folding metaphor for a wearable visualization and sonification display. *Visual Communication, 7*(3), 271–290.
8. Black, A., & Tokuda, K., (2005). The blizzard challenge 2005: Evaluating corpus-based speech synthesis on common databases. In: *Proceedings of Interspeech* (Vol. 1, pp. 77–80).
9. Boyd, D., (2006). *Friends, Friendsters, and MySpace Top 8: Writing Community into Being on Social Network Sites* (Vol. 1, pp. 7, 8).
10. Boyd, D., (2008). Why youth (heart) social network sites: The role of networked publics in teenage social life. YOUTH, IDENTITY, AND DIGITAL MEDIA, David, B., ed., The John D. and Catherine T. MacArthur Foundation Series on Digital Media and Learning, The MIT Press, Cambridge, MA, 2007-16, (Vol. 1, pp. 67–80).
11. Bradbury, R., & Gasset, J. O. Y, (2014). A sound of flower: Evolutionary teachings from complex systems. *Frontiers in Ecology, Evolution and Complexity, 1*, 193.
12. Braun, M., Mainz, A., Chadowitz, R., Pfleging, B., & Alt, F., (2019). At your service: Designing voice assistant personalities to improve

automotive user interfaces. In: *Proceedings of the 2019 CHI Conference on Human Factors in Computing Systems* (Vol. 1, pp. 1–11).

13. Brunet, M., (2021). *OER by Discipline Guide.* University of Ottawa (v. 1.0–June 2021).
14. Byeon, H. J., Lee, C., Lee, J., & Oh, U., (2022). "A voice that suits the situation": Understanding the needs and challenges for supporting end-user voice customization. In: *CHI Conference on Human Factors in Computing Systems* (Vol. 1, pp. 1–10).
15. Cambre, J., & Kulkarni, C., (2019). One voice fits all? Social implications and research challenges of designing voices for smart devices. *Proceedings of the ACM on Human-Computer Interaction, 3*(CSCW), 1–19.
16. Cambre, J., Colnago, J., Maddock, J., Tsai, J., & Kaye, J., (2020). Choice of voices: A large-scale evaluation of text-to-speech voice quality for long-form content. In: *Proceedings of the 2020 CHI Conference on Human Factors in Computing Systems* (Vol. 1, pp. 1–13).
17. Casares, Jr. D. R., (2022). Embracing the podcast era: Trends, opportunities, & implications for counselors. *Journal of Creativity in Mental Health, 17*(1), 123–138.
18. Chou, H. T. G., & Edge, N., (2012). "They are happier and having better lives than I am": The impact of using Facebook on perceptions of others' lives. *Cyberpsychology, Behavior, and Social Networking, 15*(2), 117–121.
19. Clark, L., Doyle, P., Garaialde, D., Gilmartin, E., Schlögl, S., Edlund, J., & Cowan, B. R., (2019). The state of speech in HCI: Trends, themes and challenges. *Interacting with Computers, 31*(4), 349–371.
20. Cowan, B. R., Pantidi, N., Coyle, D., Morrissey, K., Clarke, P., Al-Shehri, S., & Bandeira, N., (2017). " What can i help you with?" infrequent users' experiences of intelligent personal assistants. In: *Proceedings of the 19th International Conference on Human-Computer Interaction with Mobile Devices and Services* (Vol. 1, pp. 1–12).
21. DePaulo, B. M., (1992). Nonverbal behavior and self-presentation. *Psychological Bulletin, 111*(2), 203.
22. Doraiswamy, P. M., London, E., Varnum, P., Harvey, B., Saxena, S., Tottman, S., & Candeias, V., (2019). Empowering 8 billion minds: Enabling better mental health for all via the ethical adoption of technologies. *NAM Perspectives, 1*, 1–9.

23. Doyle, P. R., Edwards, J., Dumbleton, O., Clark, L., & Cowan, B. R., (2019). Mapping perceptions of humanness in intelligent personal assistant interaction. In: *Proceedings of the 21st International Conference on Human-Computer Interaction with Mobile Devices and Services* (Vol. 1, pp. 1–12).

24. Drager, K., (2010). Sociophonetic variation in speech perception. *Language and Linguistics Compass, 4*(7), 473–480.

25. Dubiel, M., Halvey, M., Gallegos, P. O., & King, S., (2020). Persuasive synthetic speech: Voice perception and user behavior. In: *Proceedings of the 2nd Conference on Conversational User Interfaces* (Vol. 1, pp. 1–9).

26. Eitel-Porter, R., (2021). Beyond the promise: Implementing ethical AI. *AI and Ethics, 1*(1), 73–80.

27. Elias, A., Gill, R., & Scharff, C., (2017). Aesthetic labor: Beauty politics in neoliberalism. In: *Aesthetic Labor* (Vol. 1, pp. 3–49). Palgrave Macmillan, London.

28. Ellison, N., Heino, R., & Gibbs, J., (2006). Managing impressions online: Self-presentation processes in the online dating environment. *Journal of Computer-Mediated Communication, 11*(2), 415–441.

29. Farrell, C., (2014). *Unretirement: How Baby Boomers are Changing the Way we Think About Work, Community, and the Good Life* (Vol. 1, pp. 2–9). Bloomsbury Publishing USA.

30. Fiannaca, A. J., Paradiso, A., Campbell, J., & Morris, M. R., (2018). Voice setting: Voice authoring UIs for improved expressivity in augmentative communication. In: *Proceedings of the 2018 CHI Conference on Human Factors in Computing Systems* (Vol. 1, pp. 1–12).

31. Fox, A., & Wu, J. C., (2021). Teaching modular synth & sound design online during COVID-19: Maximizing learning outcomes through open-source software and student-centered pedagogy. In: *Audio Engineering Society Convention 151* (Vol. 1, pp. 6–12). Audio Engineering Society.

32. Fröjd, M., & Horner, A., (2009). Sound texture synthesis using an overlap–add/granular synthesis approach. *Journal of the Audio Engineering Society, 57*(1, 2), 29–37.

33. Garner, T., Grimshaw, M., & Nabi, D. A., (2010). A preliminary experiment to assess the fear value of preselected sound parameters

in a survival horror game. In: *Proceedings of the 5th Audio Mostly Conference: A Conference on Interaction with Sound* (Vol. 1, pp. 1–9).

34. Goffman, E., (1978). *The Presentation of Self in Everyday Life* (Vol. 21, pp. 2–9). London: Harmondsworth.
35. Govind, D., & Prasanna, S. R., (2013). Expressive speech synthesis: A review. *International Journal of Speech Technology, 16*(2), 237–260.
36. Graps, A., (1995). An introduction to wavelets. *IEEE Computational Science and Engineering, 2*(2), 50–61.
37. Greene, B. G., Logan, J. S., & Pisoni, D. B., (1986). Perception of synthetic speech produced automatically by rule: Intelligibility of eight text-to-speech systems. *Behavior Research Methods, Instruments, & Computers, 18*(2), 100–107.
38. Halderman, J. A., Waters, B., & Felten, E. W., (2004). Privacy management for portable recording devices. In: *Proceedings of the 2004 ACM Workshop on Privacy in the Electronic Society* (Vol. 1, pp. 16–24).
39. Hogan, B., (2010). The presentation of self in the age of social media: Distinguishing performances and exhibitions online. *Bulletin of Science, Technology & Society, 30*(6), 377–386.
40. Hu, J., Xu, Q., Fu, L. P., & Xu, Y., (2019). Emojilization: An automated method for speech to emoji-labeled text. In: *Extended Abstracts of the 2019 CHI Conference on Human Factors in Computing Systems* (Vol. 1, pp. 1–6).
41. Hunt, M. G., Marx, R., Lipson, C., & Young, J., (2018). No more FOMO: Limiting social media decreases loneliness and depression. *Journal of Social and Clinical Psychology, 37*(10), 751–768.
42. Ilag, B. N., (2020). Microsoft teams overview. In: *Understanding Microsoft Teams Administration* (Vol. 1, pp. 1–36). A press, Berkeley, CA.
43. James, J., Shields, I., Berriman, R., Keegan, P. J., & Watson, C. I., (2020). Developing resources for Te Reo Māori text to speech synthesis system. In: *International Conference on Text, Speech, and Dialogue* (Vol. 1, pp. 294–302). Springer, Cham.
44. James, J., Watson, C. I., & MacDonald, B., (2018). Artificial empathy in social robots: An analysis of emotions in speech. In: *2018 27th IEEE International Symposium on Robot and Human Interactive Communication (RO-MAN)* (Vol. 1, pp. 632–637). IEEE.

45. Joinson, A. N., (2008). Looking at, looking up or keeping up with people? Motives and use of Facebook. In: *Proceedings of the SIGCHI Conference on Human Factors in Computing Systems* (Vol. 1, pp. 1027–1036).
46. Kane, S. K., Morris, M. R., Paradiso, A., & Campbell, J., (2017). " At times avuncular and cantankerous, with the reflexes of a mongoose" understanding self-expression through augmentative and alternative communication devices. In: *Proceedings of the 2017 ACM Conference on Computer Supported Cooperative Work and Social Computing* (Vol. 1, pp. 1166–1179).
47. Kim, T. H., Cho, S., Choi, S., Park, S., & Lee, S. Y., (2020). Emotional voice conversion using multitask learning with text-to-speech. In: *ICASSP 2020-2020 IEEE International Conference on Acoustics, Speech and Signal Processing (ICASSP)* (Vol. 1, pp. 7774–7778). IEEE.
48. Kobayashi, M., O'Connell, T., Gould, B., Takagi, H., & Asakawa, C., (2010). Are synthesized video descriptions acceptable?. In: *Proceedings of the 12th International ACM SIGACCESS Conference on Computers and Accessibility* (Vol. 1, pp. 163–170).
49. Koradia, Z., Aggarwal, P., Seth, A., & Luthra, G., (2013). Gurgaon idol: A singing competition over community radio and IVRS. In: *Proceedings of the 3rd ACM Symposium on Computing for Development* (Vol. 1, pp. 1–10).
50. Krämer, N. C., & Winter, S., (2008). Impression management 2.0: The relationship of self-esteem, extraversion, self-efficacy, and self-presentation within social networking sites. *Journal of Media Psychology, 20*(3), 106–116.
51. Lee, K. M., & Nass, C., (2003). Designing social presence of social actors in human computer interaction. In: *Proceedings of the SIGCHI Conference on Human Factors in Computing Systems* (Vol. 1, pp. 289–296).
52. Lee, K. M., Liao, K., & Ryu, S., (2007). Children's responses to computer-synthesized speech in educational media: Gender consistency and gender similarity effects. *Human Communication Research, 33*(3), 310–329.
53. León-Montaño, A., & Barba-Guaman, L., (2020). Design of the architecture for text recognition and reading in an online assessment applied to visually impaired students. In: *2020 International Conference*

of Digital Transformation and Innovation Technology (Incodtrin) (Vol. 1, pp. 59–65). IEEE.

54. Lindau, A., Erbes, V., Lepa, S., Maempel, H. J., Brinkman, F., & Weinzierl, S., (2014). A spatial audio quality inventory (SAQI). *Acta Acustica united with Acustica, 100*(5), 984–994.
55. Little, M., Mcsharry, P., Roberts, S., Costello, D., & Moroz, I., (2007). Exploiting nonlinear recurrence and fractal scaling properties for voice disorder detection. *Nature Precedings, 1*, 1.
56. Luger, E., & Sellen, A., (2016). " Like having a really bad pa" the gulf between user expectation and experience of conversational agents. In: *Proceedings of the 2016 CHI conference on Human Factors in Computing Systems* (Vol. 1, pp. 5286–5297).
57. Marwick, A. E., & Boyd, D., (2011). I tweet honestly, I tweet passionately: Twitter users, context collapse, and the imagined audience. *New Media & Society, 13*(1), 114–133.
58. Matias, Y., (2020). *Easier Access to Web Pages: Ask Google Assistant to Read it Aloud* (Vol. 1, pp. 1–12).
59. Mayer, R. E., Sobko, K., & Mautone, P. D., (2003). Social cues in multimedia learning: Role of speaker's voice. *Journal of educational Psychology, 95*(2), 419.
60. McGregor, M., & Tang, J. C., (2017). More to meetings: Challenges in using speech-based technology to support meetings. In: *Proceedings of the 2017 ACM Conference on Computer Supported Cooperative Work and Social Computing* (Vol. 1, pp. 2208–2220).
61. Meimaridou, E. C., Athanasopoulos, G., & Cambouropoulos, E., (2020). Musical gestures: An empirical study exploring associations between dynamically changing sound parameters of granular synthesis with hand movements. In: *14th International Symposium on Computer Music Multidisciplinary Research* (Vol. 1, p. 880).
62. Mills, T., Bunnell, H. T., & Patel, R., (2014). Towards personalized speech synthesis for augmentative and alternative communication. *Augmentative and Alternative Communication, 30*(3), 226–236.
63. Mirenda, P., Eicher, D., & Beukelman, D. R., (1989). Synthetic and natural speech preferences of male and female listeners in four age groups. *Journal of Speech, Language, and Hearing Research, 32*(1), 175–183.

64. Mori, M., MacDorman, K. F., & Kageki, N., (2012). The uncanny valley [from the field]. *IEEE Robotics & Automation Magazine, 19*(2), 98–100.

65. Mullennix, J. W., Johnson, K. A., Topcu-Durgun, M., & Farnsworth, L. M., (1995). The perceptual representation of voice gender. *The Journal of the Acoustical Society of America, 98*(6), 3080–3095.

66. Munson, B., & Babel, M., (2007). Loose lips and silver tongues, or, projecting sexual orientation through speech. *Language and Linguistics Compass, 1*(5), 416–449.

67. Niedzielski, N., (1999). The effect of social information on the perception of sociolinguistic variables. *Journal of Language and Social Psychology, 18*(1), 62–85.

68. Oosterom, N., (2021). *The Syntactic Complexity in Tracheoesophageal Speech: A Pilot Study About Complexity of Grammar in Verbal Communication After Total Laryngectomy* (Vol. 1, pp. 1–12).

69. Pearl, C., (2016). *Designing Voice User Interfaces: Principles of Conversational Experiences* (Vol. 1, pp. 4–8). " O'Reilly Media, Inc."

70. Pickup, B. A., & Thomson, S. L., (2009). Influence of asymmetric stiffness on the structural and aerodynamic response of synthetic vocal fold models. *Journal of Biomechanics, 42*(14), 2219–2225.

71. Poepel, C., (2008). Driving sound synthesis with a live audio signal. In: *Interactive Multimedia Music Technologies* (Vol. 1, pp. 167–194). IGI Global.

72. Porcheron, M., Fischer, J. E., & Sharples, S., (2017). " Do animals have accents?" Talking with agents in multi-party conversation. In: *Proceedings of the 2017 ACM Conference on Computer Supported Cooperative Work and Social Computing* (Vol. 1, pp. 207–219).

73. Pullin, G., & Hennig, S., (2015). 17 ways to say yes: Toward nuanced tone of voice in AAC and speech technology. *Augmentative and Alternative Communication, 31*(2), 170–180.

74. Purnell, T., Idsardi, W., & Baugh, J., (1999). Perceptual and phonetic experiments on American English dialect identification. *Journal of Language and Social Psychology, 18*(1), 10–30.

75. Repp, B. H., (1997). The aesthetic quality of a quantitatively average music performance: Two preliminary experiments. *Music Perception, 14*(4), 419–444.

76. Robinson, L., (2007). The cyberself: The selfing project goes online, symbolic interaction in the digital age. *New Media & Society, 9*(1), 93–110.

77. Samartzis, P., (2012). Articulating sound in a synthesized material space. In: *Supervising Practices for Postgraduate Research in Art, Architecture and Design* (Vol. 1, pp. 51–64). Brill Sense.

78. Scott, K. M., Ashby, S., Braude, D. A., & Aylett, M. P., (2019). Who owns your voice? Ethically sourced voices for non-commercial TTS applications. In: *Proceedings of the 1st International Conference on Conversational User Interface*s (Vol. 1, pp. 1–3).

79. Sebastian, R. J., & Ryan, E. B., (2018). Speech cues and social evaluation: Markers of ethnicity, social class, and age. In: *Recent Advances in Language, Communication, and Social Psychology* (Vol. 1, pp. 112–143). Routledge.

80. Sijia, X., Danač, M., Joon, S. P., Karrie, K., & Niloufar, S., (2020). Random, messy, funny, raw: Finstas as intimate reconfigurations of social media. In: *Proceedings of the 2020 CHI Conference on Human Factors in Computing Systems (Honolulu, HI, USA) (CHI '20)* (Vol. 5, pp. 1–13). Association for Computing Machinery, New York, NY, USA, Proc. ACM Hum.-Comput. Interact., No. CSCW1, Article 161. Social Media through Voice 161:21.

81. Smith, J. O., (1996). Physical modeling synthesis update. *Computer Music Journal, 20*(2), 44–56.

82. Stern, S. E., Mullennix, J. W., & Yaroslavsky, I., (2006). Persuasion and social perception of human vs. synthetic voice across person as source and computer as source conditions. *International Journal of Human-Computer Studies, 64*(1), 43–52.

83. Stern, S. E., Mullennix, J. W., Dyson, C. L., & Wilson, S. J., (1999). The persuasiveness of synthetic speech versus human speech. *Human Factors, 41*(4), 588–595.

84. Strand, E. A., (1999). Uncovering the role of gender stereotypes in speech perception. *Journal of Language and Social Psychology, 18*(1), 86–100.

85. Susini, P., McAdams, S., Winsberg, S., Perry, I., Vieillard, S., & Rodet, X., (2004). Characterizing the sound quality of air-conditioning noise. *Applied Acoustics, 65*(8), 763–790.

86. Sutton, S. J., Foulkes, P., Kirk, D., & Lawson, S., (2019). Voice as a design material: Sociophonetic inspired design strategies in human-computer interaction. In: *Proceedings of the 2019 CHI Conference on Human Factors in Computing Systems* (Vol. 1, pp. 1–14).
87. Tamagawa, R., Watson, C. I., Kuo, I. H., MacDonald, B. A., & Broadbent, E., (2011). The effects of synthesized voice accents on user perceptions of robots. *International Journal of Social Robotics, 3*(3), 253–262.
88. Tanaka, A., (2006). Interaction, experience and the future of music. In: *Consuming Music Together* (Vol. 1, pp. 267–288). Springer, Dordrecht.
89. Thomas, E. R., (2002). Sociophonetic applications of speech perception experiments. *American Speech, 77*(2), 115–147.
90. Tigwell, G. W., Gorman, B. M., & Menzies, R., (2020). Emoji accessibility for visually impaired people. In: *Proceedings of the 2020 CHI Conference on Human Factors in Computing Systems* (Vol. 1, pp. 1–14).
91. Tusing, K. J., & Dillard, J. P., (2000). The sounds of dominance. Vocal precursors of perceived dominance during interpersonal influence. *Human Communication Research, 26*(1), 148–171.
92. Valkenburg, P. M., & Peter, J., (2011). Online communication among adolescents: An integrated model of its attraction, opportunities, and risks. *Journal of Adolescent Health, 48*(2), 121–127.
93. Vashistha, A., Garg, A., Anderson, R., & Raza, A. A., (2019). Threats, abuses, flirting, and blackmail: Gender inequity in social media voice forums. In: *Proceedings of the 2019 CHI Conference on Human Factors in Computing Systems* (Vol. 1, pp. 1–13).
94. Vasquez, S., & Lewis, M., (2019). *Melnet: A Generative Model for Audio in the Frequency Domain* (Vol. 1, pp. 1–12).
95. Västfjäll, D., Gulbol, M. A., Kleiner, M., & Gärling, T., (2002). Affective evaluations of and reactions to exterior and interior vehicle auditory quality. *Journal of Sound and Vibration, 255*(3), 501–518.
96. Västfjäll, D., Kleiner, M., & Gärling, T., (2003). Affective reactions to interior aircraft sounds. *Acta Acustica united with Acustica, 89*(4), 693–701.
97. Waterloo, S. F., Baumgartner, S. E., Peter, J., & Valkenburg, P. M., (2018). Norms of online expressions of emotion: Comparing Facebook,

Twitter, Instagram, and WhatsApp. *New Media & Society, 20*(5), 1813–1831.

98. Zeng, E., Mare, S., & Roesner, F., (2017). End user security and privacy concerns with smart homes. In: *13th Symposium on Usable Privacy and Security (SOUPS 2017)* (Vol. 1, pp. 65–80).
99. Zhang, L., Jiang, L., Washington, N., Liu, A. A., Shao, J., Fourney, A., & Findlater, L., (2021). Social media through voice: Synthesized voice qualities and self-presentation. *Proceedings of the ACM on Human-Computer Interaction, 5*(CSCW1), 1–21.
100. Zhang, L., Shao, J., Liu, A. A., Jiang, L., Stangl, A., Fourney, A., & Findlater, L., (2022). Exploring interactive sound design for auditory websites. In: *CHI Conference on Human Factors in Computing Systems* (Vol. 1, pp. 1–16).
101. Zhao, X., Salehi, N., Naranjit, S., Alwaalan, S., Voida, S., & Cosley, D., (2013). The many faces of Facebook: Experiencing social media as performance, exhibition, and personal archive. In: *Proceedings of the SIGCHI Conference on Human Factors in Computing Systems* (Vol. 1, pp. 1–10).

CHAPTER 7

THE PRESENT AND FUTURE OF SOUND PROCESSING

CONTENTS

7.1. INTRODUCTION

From a physical point of view, we can state that humans can hear because tiny auditory hair cells in the inner ear detect vibrations due to sound and convert them to nerve signals. However, we also hear because, throughout evolution, the sense of hearing has helped our survival. As with many other mammals, the sense of hearing has played a major role in hunting and avoiding being hunted (Davis, 2003).

Our sense of hearing enables us to identify dangers or targets in the environment first by identifying their position in space and later by classifying them (finding out the type of animal or thing that generated the sound). The accuracy achieved by humans in these two tasks cannot be compared to any artificial development, as it is very difficult to emulate these capacities by means of computational methods (Kyriakakis et al., 1999).

Some animals outperform humans regarding their localization capacity. For example, humans can hear sounds approximately within the frequencies of 20 Hz and 20,000 Hz. Anything below 20 Hz typically cannot be heard, although it can be felt. The frequency range of dog hearing is approximately 40 Hz to 60,000 Hz, depending on the breed of dog as well as its age. Like humans, dogs can begin to go deaf as they become older. In some breeds of dogs it is natural for them to go completely deaf in their old age. In addition, dogs can move their two ears independently to improve localization accuracy (Mason et al., 2001).

Owls have a superior localization mechanism in the elevation plane, which allows them to hunt with millimetric precision. The auditory centers of the brain are responsible for interpreting the different sound signals that arrive at our two ears. These centers learn and are trained until reaching maturity. For example, babies are not able to localize sounds until they are five months old. Once these capacities are consolidated in the brain, the subject makes use of them without being aware of it. When an animal or a human being detects a hazard by a strange or uncommon sound, the brain automatically discharges in the bloodstream a load of adrenaline that warns the subject of this emergency situation (Blauert, 1997).

These involuntary actions make up the survival functions of the human auditory system. Figure 7.1 shows a map of the brain, which shows the parts associated with the processing of different types of information. Communication is a more advanced use of the sense of hearing. This use is not exclusive of human beings and it usually appears in nature among animals of the same species. We must consider that hearing is not the only mechanism used by animals for communication, as it coexists and it complements the visual, contact, tactile, and olfactory communications many times. Moreover, the different aspects of communication also vary among species depending on their sound emitting mechanisms and their cerebral development (Begault, 1999).

The sonorous language of animals is very simple and based on transmitting simple stimuli to their resemblances, for instance as a reaction to some external situation (recognizing an object, hunting, danger warning, showing love or enmity). This kind of language is generally instinctive and not learned, in contrast to articulated language, which is learned by humans. For years, scientists believed that the descent of the larynx was crucial for the development of articulated speech in humans. However, new studies show that this feature is not uniquely human, leading to new theories about the evolution of speech. These new theories state that the mechanism for the descent of the larynx is in fact composed of two phenomena that are not simultaneous: hyoid descent and larynx descent relative to the hyoid bone. Therefore, it seems that both of these anatomical moves are necessary for speech development (Macpherson and Middlebrooks, 2002).

Human beings have a third use of hearing, which is exclusive and not present in other animals. This third use of sound is considered as a means of transmission of what could be called leisure, pleasure, or art, depending on the subjective degree of sophistication that locks up the sonorous material. This third use, useless for survival and very different to the communication aim of natural language, is what is known as music. We consider music every sound expression, melodic or not, where the purpose is to entertain or satisfy the brain. The music field is without a doubt, complex, and sophisticated. It wakes up many passions in our current society and at the same time, moves an important business volume in the entertainment industry, alone or together with images, as in film or television (Algazi and Duda, 2010).

The classification of the uses of human hearing above not only gives a biological and evolutional point of view of the sense of hearing, but also can help us to understand the process of developing sound systems for any kind of purpose. If we take into account the key aspects of this description and we know how to associate them to research and business industry, our goals will always be clearer and more related to the human being they are addressed to and we will know how to integrate and take profit from these concepts to use with other disciplines in the sound industry (Wightman and Kistler, 1992).

Survival functions, such as sound localization, can give us important cues for positional sound systems and for the sound source identification that our brain is capable of achieving when important signals are merged with other noise sources. These issues will lead us to the analysis-synthesis and classification of sound disciplines. Spoken language is related to communication systems in a wide range of aspects: sound collecting, noise removal, speech enhancement, and in the last step, automatic speech recognition (ASR) (Middlebrooks, 1992).

The fact that natural language coexists with visual language, for example, is related to modern multimodal techniques, which are nowadays an important research field. Regarding music processing, we will need to incorporate advanced concepts related to artificial intelligence (AI) and machine learning. These approaches allow us to simulate how the brain performs when feelings appear as a consequence of listening to music. An example of this kind of processing are the recommendation systems that, as a function of the tastes of one person, are able to suggest similar songs. Another example are the expert systems that make possible to predict if a song will be a hit or not in the sound industry (Beitel and Kaas, 1993).

7.2. SPATIAL SOUND SYSTEMS

The objective of three-dimensional (3D) spatial sound systems is to accurately recreate the acoustic sensations that a listener would perceive inside a particular room or in an environment with certain acoustic properties. This concept implies a series of physical and technological difficulties that are a current research issue in sound engineering. Stereo sound systems, considered as the simplest approximation to spatial sound, have been utilized throughout the last 50 years as an added value in sound recordings, particularly for music material (Algazi et al., 2001).

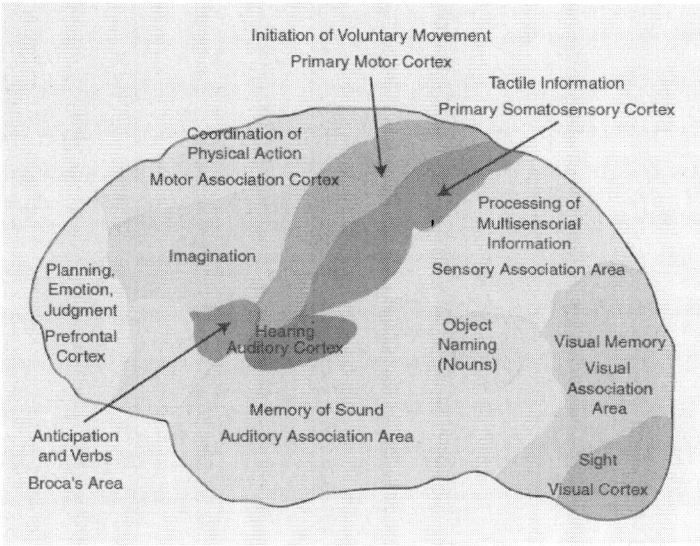

Figure 7.1. Processing centers of the brain.

Source: https://www.learnatnoon.com/s/in/briefly-describe-the-structure-of-the-following-brain/11652/?amp=1.

Used in theaters since the mid-1970s, surround sound systems have evolved and entered homes to give a better sensation than stereo by using more reproduction channels. Surround mixes are mainly intended to enhance the experience in video projections by adding artificial effects in the rear loudspeakers (explosions, reverberation, or ambient sound). The optimal listening position, known as the sweet spot, is almost limited to the central point in the loudspeaker set-up and the spatial sensation degrades considerably outside the central zone (Pralong and Carlile, 1996).

Another much more realistic strategy is to reproduce directly in the ears of the listener via headphones the signal that he would perceive in the acoustic environment to be simulated. The perception of the simulated scene depends on the fidelity of the reproduction. This strategy is widely known as binaural reproduction. The signals to be reproduced with headphones can be recorded with an acoustic head or artificially synthesized by using a measured head related transfer-function (HRTF). The future of HRTF-based techniques is promising, since a significant amount of music material is listened to over headphones using mobile devices. There are still some issues to be solved regarding the HRTF variability among different subjects and active research lines are centered on this aspect of binaural reproduction. Figure 7.2 shows a scheme of a binaural recording/reproduction system (Griesinger, 1990).

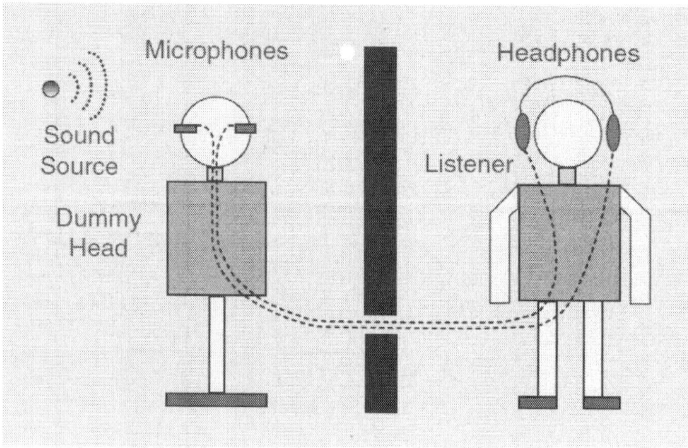

Figure 7.2. A binaural sound recording/reproduction scheme.

Source: https://www.semanticscholar.org/paper/Listen-up%E2%80%94The-present-and-future-of-audio-signal-Cobos-L%C3%B3pez/a12061c2298c0640cb15dddaaf22b3f7fa0d66b6/figure/1.

On the other hand, the most promising spatial sound system currently is called wave-field synthesis (WFS). The most basic difference of this system in comparison to 5.1 is that the acoustic field is accurately synthesized using loudspeaker arrays in a broad area, suppressing the sweet spot that characterizes conventional surround systems. WFS was first proposed by Berkhout and it is based on the two-dimensional (2D) simplification of the Huygens principle, which states that a wavefront produced by a primary source can by synthesized by a distribution of secondary sources located in the wavefront (Brungart, 1999).

The research in WFS has been very active in Europe in the last decade and several research groups are pioneers of this emerging sound system (see Figure 7.2) (Gardner, 1969).

7.3. HEADPHONE BASED SPITAL SOUND PROCESSING

Advances in communication infrastructure and technology from the cell phone to the Internet are placing us at the threshold of a new generation of mobile applications that will deliver immersive communication. Such developments will spread rapidly and will impact both the workplace and the general public (Bronkhorst and Houtgast, 1999).

This article is concerned with the generation and reproduction of spatial sound for mobile immersive communications. Properly reproduced over headphones, spatial sound can provide an astonishingly lifelike sense of being remotely immersed in the presence of people, musical instruments, and environmental sounds whose origins are either far distant, virtual, or a mixture of local, distant, and virtual. For voice communication, spatial sound can go beyond increased realism to enhancing intelligibility and can provide the natural binaural cues needed for spatial discrimination (Kendall, 1995).

Mobile voice communications applications that use these new capabilities include Sound teleconferencing or telepresence in a meeting. For music, immersive sound can go beyond reproduction that places the listener in the performance venue (or perhaps positioned on stage among the performers) to enabling the creation of entirely new Sound effects. For environmental monitoring or games, it can provide unparalleled awareness of both the sound-generating objects and the surrounding acoustic space. Spatial sound will also be used in conjunction with video in remote monitoring to provide rapid sonic detection and orientation of events for subsequent detailed analysis by video. Spatial sound technology has a long history (Wightman and Kistler, 1999).

The familiar stereo and multichannel surround-sound systems were designed for loudspeaker reproduction. By contrast, in this article, we focus on mobile systems, where the low power, light weight, high fidelity, low cost, and simple convenience of headphones make them the obvious choice. Thus, this article focuses on the generation and reproduction of headphone-based spatial sound (Wightman and Kistler, 1989).

7.3.1. Challenges

The delivery of a high-quality spatial sound experience over headphones requires reproduction of the complex dynamic signals encountered in natural hearing. This goes well beyond current commercial practice. When sound is heard over only one earphone—as is typical for contemporary cell phones—the listening experience is severely limited. A pair of earphones enables binaural reproduction, which provides a major improvement (Moore et al., 2010).

However, if a single voice channel is used to feed both earphones, most listeners will hear the voice internalized in or near the center of their heads. Relevant auditory cues can be produced by changing the balance and/or by introducing interaural time delays. These changes can shift the apparent location to a different point on a line between the ears, but the sound remains inside the head and unnatural (Møller et al., 1996).

Binaural recordings made with two microphones embedded in a dummy head introduce such basic cues as the proper interaural time and level differences and add the important acoustic cues of room reflections and reverberation. They can produce a compellingly realistic listening experience. However, because of the lack of response to head motion, there are still major problems with conventional binaural technology: a) front/back confusion (and the related failure of binaural pickup to produce externalized sound for sources that are directly in front or in back), and b) significant sensitivity to the size and shape of the listener's head and outer ears. Further, the common experience of focusing attention by turning towards the source of the sound is not possible (Kulkarni and Colburn, 1998).

As we shall explain, there are basically two different ways to exploit dynamic cues to solve these problems. One approach uses so-called head-related transfer functions (HRTFs) to filter the signals from the source in a way that accounts for the propagation of sound from the source to the listener's two ears. This approach requires having HRTFs and isolated signals for every source and uses HRTF interpolation to account for head motion. The other approach, motion-tracked binaural (MTB), is based on sampling the sound field sparsely in the space around a real or virtual dummy head. MTB requires knowing the signals at multiple points around the head and uses interpolation of the signals from these microphones to account for head motion. For both methods, the essential dynamic cues that are generated by head motion can now be achieved by lowcost, low-power, small-size head trackers based on microelectromechanical systems (MEMS) technology (Ajdler et al., 2008).

Thus, the development of new signal processing methods that respond to the dynamics of human motion promises a new era in immersive binaural Sound applications for mobile communications. Understanding any binaural technology requires knowledge of both the physics of sound propagation and the psychophysics of auditory perception. We begin with a brief review of the psychoacoustic cues for sound localization and then review their physical basis (Kistler and Wightman, 1992).

7.3.2. Sound Localization Cues

There is a large body of literature on the psychoacoustics of sound localization which can only be summarized briefly here. Blauert's book is the classic reference for the psychoacoustics of spatial sound. Begault also surveys the effects of visual and other nonauditory cues on spatial sound perception. The primary auditory cues used by people include (Duraiswaini et al., 2004)

- The interaural time difference (ITD)
- The interaural level difference (ILD)
- Monaural spectral cues that depend on the shape of the outer ear or pinna
- Cues from torso reflection and diffraction
- The ratio of direct to reverberant energy
- Cue changes induced by voluntary head motion
- Familiarity with the sound source (Haneda et al., 1999).

Except for source familiarity, all of these cues stem from the physics of sound propagation and vary with azimuth, elevation, range, and frequency. Although some of these cues are stronger than others, for optimum sound reproduction all of them should be present and consistent. When a strong cue conflicts with a weak one, the strong cue will often dominate. However, if the conflicts are too great, the listener will become bewildered, and the apparent location of the sound source will either be in error or be indeterminate (Jenison and Fissell, 1996).

The ITD and ILD are the primary cues for estimating the so-called lateral angle, the angle between the vertical median plane and a ray from the center of the head to the sound source. These cues have the important property of being largely independent of the source spectrum. According to Lord Rayleigh's well-known duplex theory, the ITD prevails at low frequencies, where head shadowing is weak, and the ILD prevails at high frequencies, where interaural phase difference is ambiguous. The crossover frequency is around 1.5 kHz, where the wavelength of sound becomes less than the distance between the ears. Subsequent research has shown that the interaural envelope delay (IED) provides a temporal localization cue at high frequencies. However, the low-frequency ITD is a particularly strong cue, and can override other, weaker localization cues (Brown and Duda, 1998).

The cues for elevation are not as robust as those for the lateral angle. It is generally accepted that the monaural spectral changes introduced by the outer ears or pinnae provide the primary static cues for elevation, although they can be overridden by head motion cues (Brungart and Scott, 2001).

These spectral changes occur above 3 kHz, where the wavelength of sound becomes smaller than the size of the pinna. The reflection and refraction of sound by the torso provides even weaker elevation cues, although they appear at lower frequencies and can be important for sources that have little high-frequency content. Monaural pinna cues present a special problem for sound reproduction because they vary so much from person to person, and they may not be faithfully reproduced by uncompensated headphones (Mackensen et al., 1999).

The three primary cues for range are the absolute loudness level combined with familiarity with the source, the low-frequency ILD for close sources, and the ratio of direct to reverberant energy for distant sources. In particular, reverberant energy decorrelates the signals reaching the two ears, and the differences between the timbre of direct and reverberant energy provides another localization cue, one that might be important for front/back discrimination as well. All of these cues contribute to externalization the sense that the origin of the sound is outside of the head (Jot, 1999).

Achieving convincing externalization with headphone-based sound reproduction has proved to be a difficult challenge, particularly for sources directly in front of or directly behind the listener. All of the cues mentioned so far are static. However, it has long been recognized that people also use dynamic cues from head motion to help localize sounds. Over 60 years ago, Wallach demonstrated that motion cues dominate pinna cues in resolving front/back confusion. Although the pinna also provides important front/back cues, and although head motion is not effective for localizing very brief sounds, subsequent research studies have confirmed the importance of dynamic cues for resolving front/back ambiguities, improving localization accuracy, and enhancing externalization (Gerzon, 1985; Johnston et al., 2019).

This summary of a large body of literature is necessarily brief, and a word of caution is needed. In particular, as is commonly done in the psychoacoustic literature, we have described the localization cues in the frequency domain, as if the ear were a Fourier spectrum analyzer. Because the auditory system performs an unusual kind of nonlinear, adaptive, short-time spectral analysis, classical spectral arguments require caution. The Franssen effect, for example, cannot be explained by a simple spectral analysis (see p. 280). The fact that multiple sound sources are almost always present further complicates spectral arguments (Davis et al., 2005).

In saying that the ITD and ILD are largely independent of the source spectrum, for example, we are tacitly assuming that the source spectrum is not changing rapidly and that there are time periods when the signal-to-noise ratio (SNR) is high across the spectrum. Despite these limitations, spectral arguments provide insight into how humans localize sounds (Algazi and Duda, 2010).

7.3.3. The HRTF, HRIR, and BRIR

The acoustic cues for sound localization are a consequence of the physical processes of sound generation, propagation, diffraction, and scattering by objects in the environment, including the listener's own body. In principle, these processes can be analyzed by solving the wave equation subject to the appropriate boundary conditions (Algazi et al., 2004). In practice, the irregularities of the boundary surfaces produce extremely complex phenomena, and measuring the boundary surfaces (particularly, the pinnae) with sufficient accuracy can be challenging. Analytical solutions are available only for very simple geometries. Standard numerical methods are limited by the need to have at least two spatial samples for the shortest wavelength of interest, and by execution times that grow as the cube of the number of sample points. Thus, most of what is known about the acoustic cues has come from acoustic measurements (Firtha et al., 2017) (Figure 7.3).

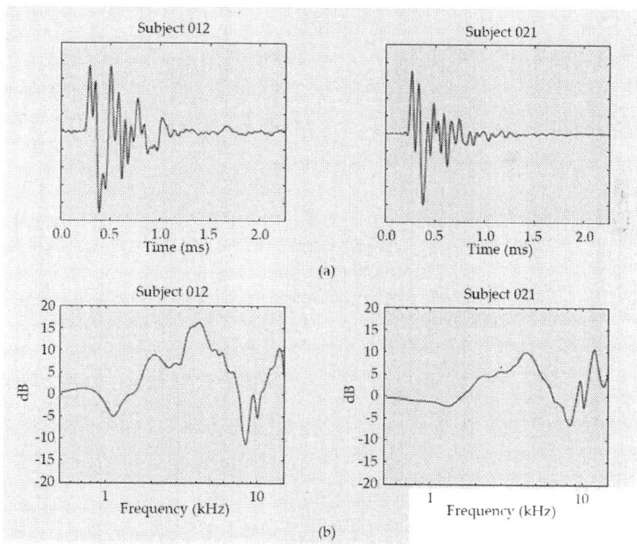

Figure 7.3. Part (a) shows the HRIRs for subject 012 and Subject 021 in the CIPIC HRTF database. Part (b) shows the magnitudes of the HRTFs.

Source: https://www.mdpi.com/2076–3417/8/11/2029/htm.

Fortunately, at typical sound pressure levels and object velocities, the physical processes are essentially linear and time invariant, and linear systems theory applies. The effects of the listener's own body on sounds coming from an isotropic point source in an anechoic environment are captured by the so called HRTF (Algazi et al., 2004; Bujacz et al., 2016). The HRTF is defined as the ratio of the Fourier transform of the sound pressure developed at the ear to the Fourier transform of the sound pressure developed at the location of the center of the listener's head with the listener absent. This frequency-domain definition has the advantage that the resulting HRTF is essentially independent of range when the source is in the far field. Most HRTF measurements are made under these conditions. The far-field range dependence is easily obtained merely by adding the propagation delay and the inverse range dependence (Algazi et al., 2005).

The inverse Fourier transform of the HRTF is the head-related impulse response (HRIR). If $h1\ t2$ is the HRIR for a distant source and c is the speed of sound, then the anechoic pressure response to an impulsive velocity source at a distance r is proportional to $h1\ t2r/c2\ /r$. The situation is more complicated when the source has a complicated radiation pattern or is distributed or is close to the head (Algazi et al., 2005)and we limit our discussion to an isotropic point source in the far-field (Kleiner et al., 1993).

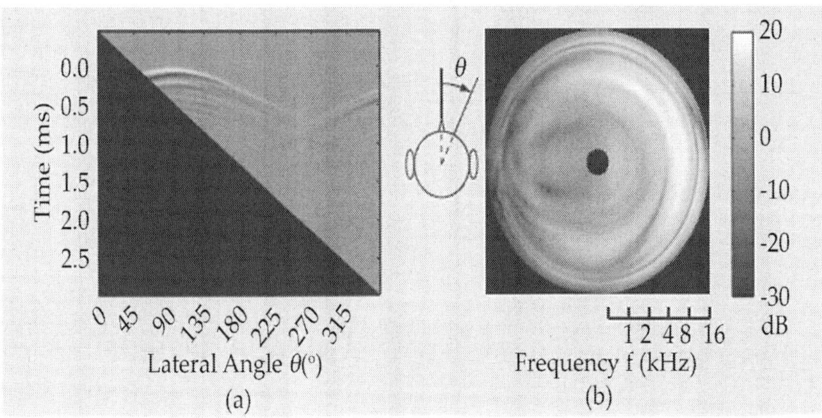

Figure 7.4. (a) Horizontal-plane variation of the right-ear HRIR and (b) the HRTF magnitude for Subject 021. In these images, the response is indicated by the brightness level.

Source: https://www.researchgate.net/Figure/HRTF-magnitudes-of-the-right-ear-in-the-horizontal-plane-at-a-distance-of-02-m-for-a_Fig4_254497582.

The temporal structure (especially multipath effects) is most easily seen in the HRIR, whereas the spectral structure is best revealed by the HRTF magnitude. Figure 7.4 shows experimentally measured HRIRs and HRTFs for two different subjects for a sound source located directly ahead. The complex behavior seen above 3 kHz is due primarily to the pinna, and the subject-to subject differences are primarily due to differences in the sizes and shapes of the subjects' pinnae. The results shown in Figures 7.4–7.5 were taken from the CIPIC HRTF database (Zölzer, 2008).

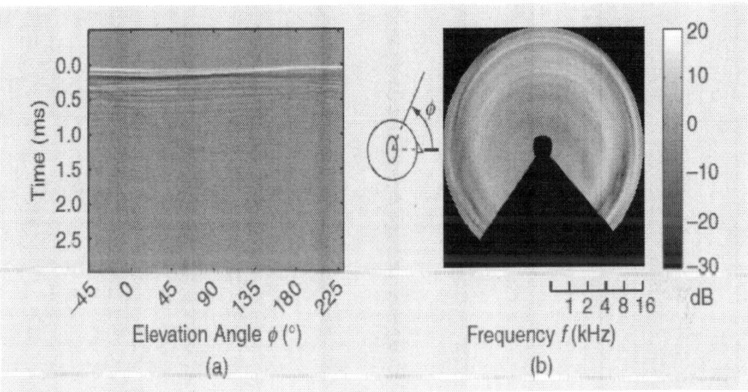

Figure 7.5. Median-plane variation of the (a) HRIR and the (b) HRTF magnitude with elevation angle f.

Source: https://www.researchgate.net/figure/a-HRIRs-b-HRTFs-of-the-right-ear-for-azimuth-angle-th-0-o-and-elevation-angles-ph_Fig3_233764855.

The directional dependence of the response for Subject 021 is illustrated in the images shown in figures shows how the right-ear HRIR changes when the source circles around the subject in the horizontal plane. The impulse response is strongest and begins soonest when the lateral angle u is close to 100° and the source is radiating directly into the ear (Huopaniemi, 2007; Belloch et al., 2013). The HRTF reveals that the magnitude response is essentially constant in all directions at low frequencies, but above 3 kHz the response on the ipsilateral side 10°, u, 180°2 is clearly greater than the response on the contralateral side 1180°, u, 360°2. To a first approximation, the response of the left ear can be found by changing the sign of u. From the plots, we see that the time of arrival and the magnitude of signals, and thus the ITD and the ILD, also vary systematically with u, and it is not surprising that the ITD and the ILD are strong cues for u (Craik, 2002).

The variation of the HRTF with the elevation angle f is more subtle. Figure 7.5 shows results in the median plane, where interaural differences are usually negligible. The HRIR reveals various pinna resonances and faint torso reflections. The HRTF shows that the strengths of the resonances and the frequencies and depths of various interference notches do change systematically with elevation. These spectral changes provide the monaural cues for elevation. The spectral profile varies significantly from person to person, and individualized HRTFs are required for accurate static elevation perception (Bernardini and De Poli, 2007) (Figure 7.6).

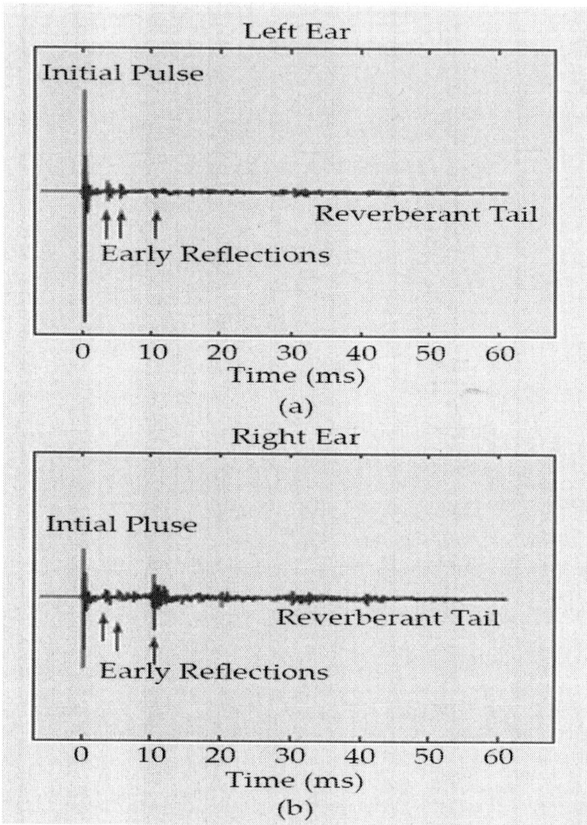

Figure 7.6. An example BRIR for a small room with the sound source on the left. The response at the left ear is shown in (a) and the response of the right ear is shown in (b). The initial pulse is the HRIR. Early reflections from the floor, ceiling, and walls are clearly visible. The multiple reflections that constitute the reverberant tail decay exponentially and last beyond the 60-ms time segment shown. Reverberation times in concert halls can extend to several seconds.

Source: https://www.nature.com/articles/s41562–021–01244-z.

In these plots, the HRTFs and HRIRs are presented as continuous functions of lateral angle and elevation. In practice, they are always sampled at discrete angles. When results are needed at intermediate angles, interpolation is required. This raises the question of how densely the HRTFs need to be sampled to achieve accurate reconstruction. The answer depends on the tolerable reconstruction error, and is ultimately a psychoacoustic question. In practice, the sampling density that is typically used is on the order of five degrees, which has received support from theoretical analysis (De Volder et al., 2013).

For practical applications as well as theoretical understanding, it is often useful to be able to replace an experimentally measured HRTF by a mathematical model. By including only a small number of terms or a small number of coefficients, these models can often be simplified or smoothed to provide HRTF approximations. Many models have been proposed, including principal components models, spherical-harmonic models, neural network models, pole-zero models, and structural models. Unfortunately, the literature is too large to be reviewed here, and the references cited only provide representative examples (Chen et al., 2008).

Listening to a sound signal filtered by individualized HRTFs produces the auditory experience of hearing that sound in an anechoic chamber. However, anechoic chambers are very unusual and unpleasant listening environments (Avendano, 2004). Although we are usually not aware of our acoustic surroundings, reflections of sound energy from objects in the environment have a profound effect on the nature and quality of the sound that we hear. In particular, for a distant source in a normal setting, the acoustic energy coming directly from the source can be significantly less than the subsequent energy arriving from multiple reflections. When the reflected sounds are missing, the perception is that the source must be very close (Rauch et al., 2006).

It is unfortunate for the developers of spatial sound systems that most people believe that they are much better at judging the distance to a sound source than they actually are. Without visual cues, people usually greatly underestimate the distance to a source from its sound alone. Interestingly, we do best when the source is a person speaking, where familiarity with the source allows us to estimate range from the loudness level (Rauch et al., 2006). In general, proper gain settings, which listeners ordinarily want to control, are important for accurate distance judgments, and this is particularly important in the case of speech (Abowd and Mynatt, 2000).

A natural way to accommodate the effects of the environment is to measure the impulse response in a room, thereby including all of the early reflections and subsequent reverberation caused by multiple reflections. When separate measurements are made for each ear, this is called the binaural room impulse response (BRIR). As Figure 7.4 illustrates, BRIRs are much longer than HRIRs. Thus, in filtering a sound signal with BRIRs, the issues of latency and computation time must be addressed (Lloyd-Fox et al., 2010).

7.4. ANALYSIS, CLASSIFICATION, AND SEPARATION OF SOUNDS

One of the most interesting research lines is the sound processing oriented to intelligent segmentation, classification, or separation of the auditory events present in Sound signals. Researchers have tried to give a solution to the well-known cocktail party problem throughout decades. The cocktail party effect describes the ability to focus one's listening attention on a single talker among a mixture of conversations and background noises, ignoring other conversations (Stryker and Burke, 2000) (Figure 7.7).

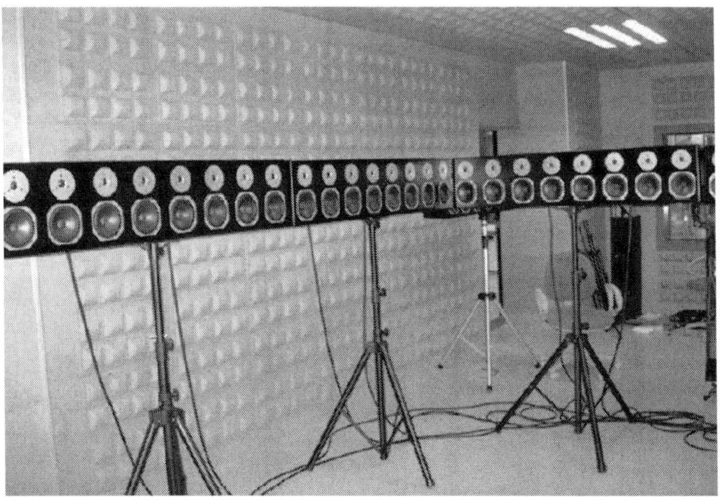

Figure 7.7. Wave-field synthesis prototype.

Source: http://personales.upv.es/jjlopez/WFS.html.

This effect reveals one of the surprising abilities of our auditory system, which enables us to talk in a noisy place. It also reveals how our auditory system and brain are capable of combining themselves in a unique way (Breebaart et al., 2004). The task of separating a signal from a mixture is not easy and there is no machine capable of performing with the accuracy achieved by humans. The traditional approaches such as adaptive beamforming or independent component analysis (ICA), provide a solution by using several sensors and statistical analysis, but they are not as effective as our auditory system. That is why, in contrast to ICA and other purely statistical techniques, there is another discipline known as computational auditory scene analysis (CASA) (Drennan et al., 2014).

CASA methods try to emulate the human auditory system by analyzing the sound in small time windows by using bandpass filter banks that simulate our own auditory perception. These filter banks provide a kind of time-frequency decomposition which is a common step in sound classification and separation. Many separation algorithms are based on grouping time-frequency elements resulting from a time-frequency representation (Mollah et al., 2004). There are many open questions such as, what is the most convenient time frequency representation for having a good separation quality or what are the grouping cues that lead to better results? An active research line regarding these questions is the one that studies sparse methods for sound separation. If a signal is sparse under a given representation, its information is concentrated in few elements and separating sparse signals from a mixture becomes an easier task (Alexandridis et al., 2013).

Sound source separation has a lot of potential applications including:

Sound up-

- mixing (Miller, 1994):

High quality Sound separation would make possible to adapt mono and stereo material to present and future spatial sound systems, as 5.1 and WFS.

- Music remixing and reprocessing: Extracting a sound source from a mixture can be used for karaoke applications or intelligent/ selective reprocessing of music.
- Speech enhancement: Separating concurrent speech sources can be used for improving intelligibility in speech communication systems, such as videoconferencing or free-hand devices.
- Other applications (Weinstein and Klein, 1996): Generally, carrying out a separation of the sources as a previous step for other applications can improve the performance of other systems that will be next described, for example, ASR systems or music information retrieval (MIR) tools.

Figure 7.8 shows a separation/classification scheme, where the Sound input from a stereo CD is analyzed for separation and identification the different instruments that are playing (Osterwalder et al., 2005).

7.5. AUTOMATIC SPEECH RECOGNITION AND SYNTHESIS

Recognizing speech from a given speaker, with previous training and under very controlled situations, is a completely feasible task. There are also ASR systems independent from the speaker, but they are limited to a small vocabulary intended for a simple situation, for example, those used in the automatic selling of tickets for travel agencies or shows (Niemczynowicz, 1999).

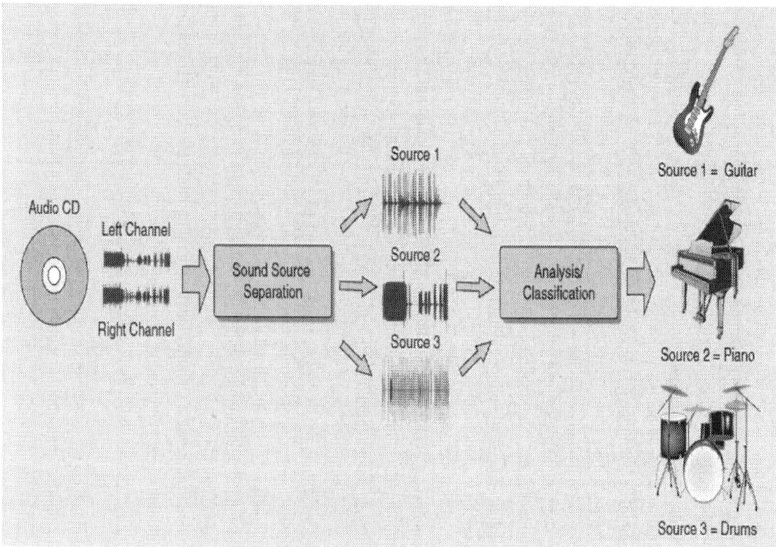

Figure 7.8. A sound source separation/classification scheme.

Source: https://dcase.community/challenge2020/task-sound-event-detection-and-separation-in-domestic-environments.

However, the global ASR problem is still far from being solved. Humans are not capable of understanding words and sentences in every situation with 100% accuracy. Noise, inattention, bad pronunciation, or a peculiar accent would lead us to fail in the recognition task. Sometimes we do not have the time to listen to all of the spoken words, but we extrapolate by analyzing the context (Murray et al., 2001).

Sometimes we read lips or simply ask the other person to repeat his words. By analyzing these situations, researchers have begun to incorporate other techniques in ASR systems that are closely related to this human behavior. These techniques are known as multimodal approaches and are raising the interest of the scientific community (Lee, 2015). Another aspect related to ASR systems is the collecting of sound signals in hostile environments. Regarding this aspect, microphone arrays provide a powerful solution that outperforms the auditory system, which has only two ears. Several advances have been observed in the last few years regarding speech synthesis, the sentence prosody of one of the multidisciplinary aspects that has yet be solved. Commercial systems such as Loquendo provide very good results with the aid of big databases with numerous articulations. In the field of the singing voice, Vocaloid has been a revolution (Scheirman, 2015).

7.6. SOUND COMPRESSION

7.6.1. Beginning and Evolution

Sound compression, also known as Sound coding, is a technique that employs two basic principles to reduce the number of bits that are necessary for archiving or transmitting a fragment of digital Sound. On one hand, lossless Sound compression techniques are only based on the reduction of signal redundancies (Krauss and Richard, 1999).

The signal resultant from the compression/decompression process is unaltered, but the compression rate is not higher than 2:1 for CD standard sampling. This upper bound can only be exceeded if we take profit of lossy compression techniques, where the resultant signal is different from the original but almost identical from the perceptual point of view for an average listener in the usual listening conditions (Fleck et al., 2010).

Although these techniques have been investigated since the 1970s, it was not until the appearance of the MP3 when their use became global. The history of MP3 is an example of a failure and a success of the telecommunication technologies industry. During the development of the digital radio (DAB), the European Union helped finance a project with the purpose of developing the necessary technology for DAB. One of the tasks of the project was to develop compressing efficient algorithms to be used in DAB (Cadena et al., 2016).

As a result, a compressor that was soon included in the MPEG-1 standard was developed, comprising its layer III. Finally, the MPEG-1 Layer III was discarded to be used in the DAB due to its high computational cost and its required computing time. Eventually, DAB did not take off and it is considered a great failure since, 20 years later, conventional FM radio is still by far most preferred by audiences (Wynn et al., 2014).

Nevertheless, the results obtained with Layer III were very good and as soon as personal computers had enough power to decompress Layer III in real time, the great expansion of this standard was a reality. The first decoders for PCs arose in 1997 (AMP, WinAmp). The file extension.mp3 became very popular thanks to the use of the Internet, which allowed an easy interchange of songs. P2P programs and online music shops also became very popular and portable devices were able to reproduce MP3 from SDRAM memories or tiny hard disks that flooded the market (Pérez-Cañado, 2012).

7.6.2. Present and Future

After MP3, other standards of Sound compression appeared including advance Sound coding (AAC), Dolby Advanced Codec 3 (AC3), Digital Theater System (DTS), adaptive transform acoustic coding (ATRAC), and OGG. Except for OGG, the rest are property of different consumer electronics companies and almost all of them allow coding more than two channels, going from stereo to surround sound (5.1, 6.1, 7.1). We can consider that the most advanced coder at the moment is the AAC in its improved version, in particular the HE-AAC v2, which uses spectral band replication techniques and parametric stereo, obtaining a rate of only 24 bits of Kbps for a stereo music signal (Culbertson et al., 2018).

The limits of lossy compression have been practically reached because almost all we currently know about how our auditory system discards unuseful information has been translated to compression algorithms. A new research line directly related to the compression of information is compressed sensing. The idea is to make sensors capable of directly obtaining the relevant information of the signal to be sampled and to avoid the need for sampling at the maximum rate and discarding redundant information in a second step. In addition, the concept of spatial Sound coding (SAC) is also being addressed in the compression community (Bunge, 1960).

The aim of SAC is to code spatial information in a more efficient way, reducing the rate of 5.1 surround signals. The MPEG-Surround standard is based on SAC techniques. Research in Sound compression is moving toward other aspects related to the spatiality of sound rather than achieving higher compression rates in the traditional sense, since the capacity of memory devices is continuously growing and the need for saving space is slowly vanishing (Algazi and Duda, 2010).

7.7. EXPERT SYSTEMS

The world of music processing and music databases is being revolutioned by expert systems. The concept of MIR includes all the systems that are capable of extracting and describing information from an Sound file in a similar way that humans would do it: by generating semantic data associated to the complex human perception of music (Belloch et al., 2012).

7.7.1. Recommendation Systems

Recommendation systems have been receiving increased attention in the Sound world for their potential commercial applications. A personalized music program can be automatically designed by means of specific software that takes into account the musical preferences of the listener (Narbutt et al., 2020). Tastes are usually related to music styles or some music features. A categorization of the songs according to different aspects must be carried out, and different features related to the pitch, tempo, energy distributions, or timbre are usually needed. There are some Web sites that automatically provide music recommendations that are not based on the Sound content but in tags introduced by experts (Spagnol et al., 2018).

Some interesting results have been obtained by commercial systems that combine tags with automatically extracted features. Nevertheless, an automatic classification superior to human personnel is rather impossible, as there is not even an agreement in what makes some music styles different from others (Villegas and Cohen, 2010).

7.7.2. Hit Prediction Systems

One more step in recommendation systems are the applications that are able to predict the success of a song before launching it to the market. In this case, the features looked for in a song are related to the tastes of the average audience, attempting to find those that are common and possess the increased potential of becoming a hit (Kendall et al., 1990). It is difficult to find scientific publications in this field and visible results are often provided by companies that commercialize these services (i.e., uPlaya or EchoNest, which was launched by researchers from the Massachusetts Institute of Technology). The economic consequences of the use of these systems can be very high, especially when the commercial side of music is prioritized as opposed to other artistic qualities (Gutierrez-Parera and Lopez, 2016).

7.8. CONCLUSION

Throughout this article, we have given an overview of the different research areas in the Sound processing community. Current Sound processing techniques are becoming oriented to the way humans process and perceive sound. In the future, statistical analysis, databases, the growing computation capacity, multimodal techniques, and neurosciences will play a major role in the research and development of new algorithms and systems related to sound and hearing.

REFERENCES

1. Abowd, G. D., & Mynatt, E. D., (2000). Charting past, present, and future research in ubiquitous computing. *ACM Transactions on Computer-Human Interaction (TOCHI), 7*(1), 29–58.
2. Ajdler, T., Faller, C., Sbaiz, L., & Vetterli, M., (2008). Sound field analysis along a circle and its application to HRTF interpolation. *Journal of the Audio Engineering Society, 56*(article), Vol. 1, 156–175.
3. Alexandridis, A., Griffin, A., & Mouchtaris, A., (2013). Capturing and reproducing spatial audio based on a circular microphone array. *Journal of Electrical and Computer Engineering, 2013.*
4. Algazi, V. R., & Duda, R. O., (2005). Immersive spatial sound for mobile multimedia. In: *Seventh IEEE International Symposium on Multimedia (ISM'05)* (Vol. 1, p. 8). IEEE.
5. Algazi, V. R., & Duda, R. O., (2010). Headphone-based spatial sound. *IEEE Signal Processing Magazine, 28*(1), 33–42.
6. Algazi, V. R., Avendano, C., & Duda, R. O., (2001). Elevation localization and head-related transfer function analysis at low frequencies. *The Journal of the Acoustical Society of America, 109*(3), 1110–1122.
7. Algazi, V. R., Duda, R. O., & Thompson, D. M., (2004). Motion-tracked binaural sound. *Journal of the Audio Engineering Society, 52*(11), 1142–1156.
8. Algazi, V. R., Duda, R. O., Melick, J. B., & Thompson, D. M., (2004). Customization for personalized rendering of motion-tracked binaural sound. In*: Audio Engineering Society Convention 117: Audio Engineering Society* (Vol. 1, pp. 2–9).
9. Avendano, C., (2004). Virtual spatial sound. In: *Audio Signal Processing for Next-Generation Multimedia Communication Systems* (Vol. 1, pp. 345–370). Springer, Boston, MA.
10. Begault, D. R., (1999). Auditory and non-auditory factors that potentially influence virtual acoustic imagery. In: *Audio Engineering Society Conference: 16th International Conference: Spatial Sound Reproduction. Audio Engineering Society*, (Vol. 1, pp. 2–9).
11. Beitel, R. E., & Kaas, J. H., (1993). Effects of bilateral and unilateral ablation of auditory cortex in cats on the unconditioned head orienting response to acoustic stimuli. *Journal of Neurophysiology, 70*(1), 351–369.

12. Belloch, J. A., Ferrer, M., Gonzalez, A., Martinez-Zaldivar, F. J., & Vidal, A. M., (2013). Headphone-based virtual spatialization of sound with a GPU accelerator. *Journal of the Audio Engineering Society, 61*(7, 8), 546–561.
13. Belloch, J. A., Ferrer, M., Gonzalez, A., Martínez-Zaldívar, F. J., & Vidal, A. M., (2012). Headphone-based spatial sound with a GPU accelerator. *Procedia Computer Science, 9*(1), 116–125.
14. Bernardini, N., & De Poli, G., (2007). The sound and music computing field: Present and future. *Journal of New Music Research, 36*(3), 143–148.
15. Blauert, J., (1997). *Spatial Hearing: The Psychophysics of Human Sound Localization*. MIT press.
16. Breebaart, J., Van De, P. S., Kohlrausch, A., & Schuijers, E., (2004). High-quality parametric spatial audio coding at low bitrates. In: *Audio Engineering Society Convention 116: Audio Engineering Society,* (Vol. 1, pp. 2–9).
17. Bronkhorst, A. W., & Houtgast, T., (1999). Auditory distance perception in rooms. *Nature, 397*(6719), 517–520.
18. Brown, C. P., & Duda, R. O., (1998). A structural model for binaural sound synthesis. *IEEE Transactions on Speech and Audio Processing, 6*(5), 476–488.
19. Brungart, D. S., & Scott, K. R., (2001). The effects of production and presentation level on the auditory distance perception of speech. *The Journal of the Acoustical Society of America, 110*(1), 425–440.
20. Brungart, D. S., (1999). Auditory localization of nearby sources. III. Stimulus effects. *The Journal of the Acoustical Society of America, 106*(6), 3589–3602.
21. Bujacz, M., Kropidlowski, K., Ivanica, G., Moldoveanu, A., Saitis, C., Csapo, A., & Witek, P., (2016). Sound of Vision-Spatial audio output and sonification approaches. In: *International Conference on Computers Helping People with Special Needs* (Vol. 1, pp. 202–209). Springer, Cham.
22. Bunge, W., (1960). *The Economic Base of the Puget Sound Region, Present and Future* (Vol. 1, pp. 4–9). Washington State Department of Commerce and Economic Development, Business and Economic Research Division.

23. Cadena, C., Carlone, L., Carrillo, H., Latif, Y., Scaramuzza, D., Neira, J., & Leonard, J. J., (2016). Past, present, and future of simultaneous localization and mapping: Toward the robust-perception age. *IEEE Transactions on Robotics, 32*(6), 1309–1332.
24. Chen, D., Doumeingts, G., & Vernadat, F., (2008). Architectures for enterprise integration and interoperability: Past, present and future. *Computers in Industry, 59*(7), 647–659.
25. Craik, F. I., (2002). Levels of processing: Past, present.. and future?. *Memory, 10*(5, 6), 305–318.
26. Culbertson, H., Schorr, S. B., & Okamura, A. M., (2018). Haptics: The present and future of artificial touch sensation. *Annual Review of Control, Robotics, and Autonomous Systems, 1*(1), 385–409.
27. Davis, L. S., Duraiswami, R., Grassi, E., Gumerov, N. A., Li, Z., & Zotkin, D. N., (2005). High order spatial audio capture and its binaural head-tracked playback over headphones with HRTF cues. In: *Audio Engineering Society Convention 119: Audio Engineering Society,* (Vol. 1, pp. 4–8).
28. Davis, M. F., (2003). History of spatial coding. *Journal of the Audio Engineering Society, 51*(6), 554–569.
29. De Volder, M. F., Tawfick, S. H., Baughman, R. H., & Hart, A. J., (2013). Carbon nanotubes: Present and future commercial applications. *Science, 339*(6119), 535–539.
30. Drennan, W. R., Svirsky, M. A., Fitzgerald, M. B., & Rubinstein, J. T., (2014). Mimicking normal auditory functions with cochlear implant sound processing; past, present and future. In: Susan, B. W., & Thomas, R. J., (eds), *Cochlear Implants* (Vol. 1, pp. 3–8). Thieme, New-York.
31. Duraiswaini, R., Zotkin, D. N., & Gumerov, N. A., (2004). Interpolation and range extrapolation of HRTFs [head related transfer functions]. In: *2004 IEEE International Conference on Acoustics, Speech, and Signal Processing* (Vol. 4, pp. iv). IEEE.
32. Firtha, G., Fiala, P., Schultz, F., & Spors, S., (2017). Improved referencing schemes for 2.5 D wave field synthesis driving functions. *IEEE/ACM Transactions on Audio, Speech, and Language Processing, 25*(5), 1117–1127.
33. Fleck, N. A., Deshpande, V. S., & Ashby, M. F., (2010). Micro-architectured materials: Past, present and future. *Proceedings of the*

Royal Society A: Mathematical, Physical and Engineering Sciences, 466(2121), 2495–2516.

34. Gardner, M. B., (1969). Distance estimation of 0 or apparent 0-oriented speech signals in anechoic space. *The Journal of the Acoustical Society of America, 45*(1), 47–53.
35. Gerzon, M. A., (1985). Ambisonics in multichannel broadcasting and video. *Journal of the Audio Engineering Society, 33*(11), 859–871.
36. Griesinger, D., (1990). Binaural techniques for music reproduction. In: *Audio Engineering Society Conference: 8th International Conference: The Sound of Audio: Audio Engineering Society,* (Vol. 1, pp. 4–8).
37. Gutierrez-Parera, P., & Lopez, J. J., (2016). Influence of the quality of consumer headphones in the perception of spatial audio. *Applied Sciences, 6*(4), 117.
38. Haneda, Y., Makino, S., Kaneda, Y., & Kitawaki, N., (1999). Common-acoustical-pole and zero modeling of head-related transfer functions. *IEEE Transactions on Speech and Audio Processing, 7*(2), 188–196.
39. Huopaniemi, J., (2007). Future of personal audio: Smart applications and immersive communication. In: *Audio Engineering Society Conference: 30th International Conference: Intelligent Audio Environments: Audio Engineering Society* (Vol. 1, pp. 4–9).
40. Jenison, R. L., & Fissell, K., (1996). A spherical basis function neural network for modeling auditory space. *Neural Computation, 8*(1), 115–128.
41. Johnston, D., Egermann, H., & Kearney, G., (2019). Measuring the behavioral response to spatial audio within a multi-modal virtual reality environment in children with autism spectrum disorder. *Applied Sciences, 9*(15), 3152.
42. Jot, J. M., (1999). Real-time spatial processing of sounds for music, multimedia and interactive human-computer interfaces. *Multimedia Systems, 7*(1), 55–69.
43. Kendall, G. S., (1995). The decorrelation of audio signals and its impact on spatial imagery. *Computer Music Journal, 19*(4), 71–87.
44. Kendall, G. S., Martens, W. L., & Wilde, M. D., (1990). A spatial sound processor for loudspeaker and headphone reproduction. In: *Audio Engineering Society Conference: 8th International Conference: The Sound of Audio: Audio Engineering Society* (Vol. 1, pp. 4–9).

45. Kistler, D. J., & Wightman, F. L., (1992). A model of head-related transfer functions based on principal components analysis and minimum-phase reconstruction. *The Journal of the Acoustical Society of America, 91*(3), 1637–1647.

46. Kleiner, M., Dalenbäck, B. I., & Svensson, P., (1993). Auralization: an overview. *Journal of the Audio Engineering Society, 41*(11), 861–875.

47. Krauss, T. F., & Richard, M. D. L. R., (1999). Photonic crystals in the optical regime—past, present and future. *Progress in Quantum electronics, 23*(2), 51–96.

48. Kulkarni, A., & Colburn, H. S., (1998). Role of spectral detail in sound-source localization. *Nature, 396*(6713), 747–749.

49. Kyriakakis, C., Tsakalides, P., & Holman, T., (1999). Surrounded by sound. *IEEE Signal Processing Magazine, 16*(1), 55–66.

50. Lee, E. A., (2015). The past, present and future of cyber-physical systems: A focus on models. *Sensors, 15*(3), 4837–4869.

51. Lloyd-Fox, S., Blasi, A., & Elwell, C. E., (2010). Illuminating the developing brain: The past, present and future of functional near infrared spectroscopy. *Neuroscience & Biobehavioral Reviews, 34*(3), 269–284.

52. Mackensen, P., Felderhoff, U., Theile, G., Horbach, U., & Pellegrini, R. S., (1999). Binaural room scanning-a new tool for acoustic and psychoacoustic research. *The Journal of the Acoustical Society of America, 105*(2), 1343, 1344.

53. Macpherson, E. A., & Middlebrooks, J. C., (2002). Listener weighting of cues for lateral angle: The duplex theory of sound localization revisited. *The Journal of the Acoustical Society of America, 111*(5), 2219–2236.

54. Mason, R., Ford, N., Rumsey, F., & De Bruyn, B., (2001). Verbal and nonverbal elicitation techniques in the subjective assessment of spatial sound reproduction. *Journal of the Audio Engineering Society, 49*(5), 366–384.

55. Middlebrooks, J. C., (1992). Narrow-band sound localization related to external ear acoustics. *The Journal of the Acoustical Society of America, 92*(5), 2607–2624.

56. Miller, R. A., (1994). Medical diagnostic decision support systems—Past, present, and future: A threaded bibliography and brief commentary. *Journal of the American Medical Informatics Association, 1*(1), 8–27.

57. Mollah, M. Y., Morkovsky, P., Gomes, J. A., Kesmez, M., Parga, J., & Cocke, D. L., (2004). Fundamentals, present and future perspectives of electrocoagulation. *Journal of Hazardous Materials, 114*(1–3), 199–210.
58. Møller, H., Sørensen, M. F., Jensen, C. B., & Hammershøi, D., (1996). Binaural technique: Do we need individual recordings?. *Journal of the Audio Engineering Society, 44*(6), 451–469.
59. Moore, A. H., Tew, A. I., & Nicol, R., (2010). An initial validation of individualized crosstalk cancellation filters for binaural perceptual experiments. *Journal of the Audio Engineering Society, 58*(1, 2), 36–45.
60. Murray, J. M., Delahunty, C. M., & Baxter, I. A., (2001). Descriptive sensory analysis: Past, present and future. *Food Research International, 34*(6), 461–471.
61. Narbutt, M., Skoglund, J., Allen, A., Chinen, M., Barry, D., & Hines, A., (2020). Ambiqual: Towards a quality metric for headphone rendered compressed ambisonic spatial audio. *Applied Sciences, 10*(9), 3188.
62. Niemczynowicz, J., (1999). Urban hydrology and water management–present and future challenges. *Urban Water, 1*(1), 1–14.
63. Osterwalder, A., Pigneur, Y., & Tucci, C. L., (2005). Clarifying business models: Origins, present, and future of the concept. *Communications of the association for Information Systems, 16*(1), 1–7.
64. Pérez-Cañado, M. L., (2012). CLIL research in Europe: Past, present, and future. *International Journal of Bilingual Education and Bilingualism, 15*(3), 315–341.
65. Pralong, D., & Carlile, S., (1996). The role of individualized headphone calibration for the generation of high fidelity virtual auditory space. *The Journal of the Acoustical Society of America, 100*(6), 3785–3793.
66. Rauch, S. L., Shin, L. M., & Phelps, E. A., (2006). Neurocircuitry models of posttraumatic stress disorder and extinction: Human neuroimaging research—Past, present, and future. *Biological Psychiatry, 60*(4), 376–382.
67. Scheirman, D., (2015). Large-scale loudspeaker arrays: Past, present and future (part two—electroacoustic considerations). In: *Audio Engineering Society Conference: 59th International Conference: Sound Reinforcement Engineering and Technology. Audio Engineering Society,* (Vol. 1, pp. 4–9).

68. Spagnol, S., Wersényi, G., Bujacz, M., Bălan, O., Herrera, M. M., Moldoveanu, A., & Unnthorsson, R., (2018). Current use and future perspectives of spatial audio technologies in electronic travel aids. *Wireless Communications and Mobile Computing*, 1, 2–9.

69. Stryker, S., & Burke, P. J., (2000). The past, present, and future of an identity theory. *Social Psychology Quarterly, 2*(1) 284–297.

70. Villegas, J., & Cohen, M., (2010). HRIR~ modulating range in headphone-reproduced spatial audio. In: *Proceedings of the 9th ACM SIGGRAPH Conference on Virtual-Reality Continuum and its Applications in Industry* (Vol. 1, pp. 89–94).

71. Weinstein, N. D., & Klein, W. M., (1996). Unrealistic optimism: Present and future. *Journal of Social and Clinical Psychology, 15*(1), 1–8.

72. Wightman, F. L., & Kistler, D. J., (1989). Headphone simulation of free-field listening. I: Stimulus synthesis. *The Journal of the Acoustical Society of America, 85*(2), 858–867.

73. Wightman, F. L., & Kistler, D. J., (1992). The dominant role of low-frequency interaural time differences in sound localization. *The Journal of the Acoustical Society of America, 91*(3), 1648–1661.

74. Wightman, F. L., & Kistler, D. J., (1999). Resolution of front–back ambiguity in spatial hearing by listener and source movement. *The Journal of the Acoustical Society of America, 105*(5), 2841–2853.

75. Wynn, R. B., Huvenne, V. A., Le Bas, T. P., Murton, B. J., Connelly, D. P., Bett, B. J., & Hunt, J. E., (2014). Autonomous underwater vehicles (AUVs): Their past, present and future contributions to the advancement of marine geoscience. *Marine Geology, 352*(1), 451–468.

76. Zölzer, U., (2008). *Digital Audio Signal Processing* (Vol. 1, pp. 5–9). John Wiley & Sons.

CHAPTER 8

MOBILE SOUND PROCESSING FOR SOCIAL MEDIA

CONTENTS

8.1. INTRODUCTION

Social media is a vital tool for companies and people to exchange information and interact. However, since individuals are engaged across several platforms, it may be tough to break thru the clutter and engage with them. After all, the number of social media means you have the potential to connect with other people if you really can persuade consumers to respond to your messages (Shneiderman and Plaisant, 2010; Quan-Haase and Sloan, 2017).

Including audio material on your website might be just what you need to get your message heard. Instead of relying just on sight, sound allows you to communicate with consumers via two of their five sensory systems (Bergeron, 2005; Safko, 2010).). And, given that sound makes up almost 50% of the consumer experience, it is a fantastic approach to increase social media audience interaction.

It is not news that advertising is increasingly moving away from the previous press in favor of new internet forms. The usage of mobile devices to distribute sponsored material is on the rise, gaining over social media and opening up new avenues for attracting consumers' interest (Bevilacqua et al., 2011). Academic studies and public marketing reports unequivocally support the good impression of graphical material when combined with auditory data. Video-based advertising styles are becoming more prevalent(Turner and Shah, 2014). According to the most current Facebook data, streaming video receives over 8 billion per day video ads on the networking site. Furthermore, videos have the greatest engagement levels, and Facebook predicts that by 2020, the video would account for over 75% of global mobile data traffic (WordStream, 2018). The impact of the transmitted material may therefore be substantially affected by the performance of the specific pieces wherein the overall video product is constructed, not only on social networking sites as well as on other streaming services. The audio track is among the video's most identifiable sub-objects. The backing clip is often a blend of the spoken word, accompanying music, and sound effects that secures and transmits the final atmosphere and imparts sensations and emotions to the listeners.

Companies are being asked to stay up along the quick market structure, which is defined by the growth of social networking fined by the velocity through which material is spread across the society. Furthermore, as a result of this occurrence, the structure of marketing techniques has shifted considerably, from well-planned and scheduled long-term campaigns to fast-changing, almost dramatic brief attempts(Hansen et al, 2010). Customers want to be provided with material that is both innovative and natural while also reflecting a high degree of quality. It's tough to achieve all of these standards at the same time. As a result, discovering solutions that might assist speed up the manufacturing process while maintaining quality could be a huge marketing benefit.

In just this lesson, we'll look at how wearable devices (phones and tablets) may be used to replace conventional sound manufacturing equipment. The test would compare audio waves captured using a typical, complicated, and rising record chain alongside signals collected that use less sophisticated tools such as a smartphone or tablet(Lian, 2008; Junco, 2014). The major goal of this experiment is to see if it is feasible to produce high-quality results with minimal time and with less complex technology.

8.2. TECHNOLOGICAL ADVANCEMENTS IN THE MOBILE SOUND PROCESSING

Computer technology is continually growing and even accelerating the process in which new ideas are presented to the marketplace due to technological advancements. Computers and the necessary software programs are becoming a vital element of almost all sectors, not just the technology sector. Thus, it is not surprising that the field of sound transmission is also experiencing significant developments. The implementation of a digital system and hardware components in use in audio handling never ceases, and customers are regularly provided with software and hardware, as well as software having expanded characteristics and attributes (Van Till, 2018).

Currently, a range of devices capable of adjusting and processing input audio data is accessible on the market. To get a high-quality record of the original sound source, it is essential not to undervalue some of the next recording chain hardware components. The microphone is the first input. The choice of the microphone has a considerable influence on the effectiveness of the sound recording. The challenge of selecting an appropriate microphone kind for a certain use is a vast one. There is no global microphone capable of meeting various recording needs. The selection of a particular microphone kind is constantly influenced by the portfolio of accessible microphones, the sound resource, and the record-setting (Aumond et al., 2017).

The microphone preamplifier is an additional component in the recording loop. Even a single component may significantly alter the final audio level. There is a multitude of analog microphone amplification on the market.

There are recognized and tried-and-true models that have been applied to famous and praised records and are already in usage. The pricing range of the many kinds and models may be rather extensive, and the cost of certain models is often quite expensive. Microphone amplifiers, which may also be a component of sound cards, represent the less expensive choice. Built-in microphone amplifiers often cannot match with separate hold equipment in terms of quality (Biswas et al., 2017).

The third component of the circuit is the AD/DA converters, which transform the analog–to–digital version that is saved on the single disk in the right format. Additionally, an AD/DA conversion may be incorporated inside the sound card. The performance of sound cards varies greatly from kind to type. If we begin with the lowest quality category, we must name a few of the most prominent makers and manufacturers. These businesses consist of M-Audio, ESI, and LEXICON(Manjunath et al., 2002; Brandstein and Ward, 2001).

Echo Audio, Presonus, Akai, and Native Instrument are some of the firms that manufacture sound cards of good value. Nowadays, almost each DAW software provider creates and offers a full bundle that includes the program and the necessary hardware, such as a sound system or MIDI hardware driver(Pejovic et al., 2016). An alternative option is to purchase items that are the result of tight cooperation among hardware and software makers. RME, Lynx, Solid State Logic, Benchmark, and Lavry are some of the top AD/DA converters (Silva et al., 2009).

There are also other effective initiatives to transform such hardware devices effectively into the computerized environment via their software virtualization, made possible by the increasing behavior of selected computer parts as well as software products. Successful software simulations of physical components may be incorporated as VST instruments or plug-ins in the DAW software application(Vaseghi, 2008; Fujimoto, 2001). Several plug-ins are built out of their hardware as well, and as a result, they neither slow down the CPU nor impact the operating capability of the machine upon which DAW is operating. Universal Audio is among the financially successful firms using this strategy. They provide sound cards which could be used as a substitute for preamplifiers from Neve, UA, Fairchild, Manley, API, and others. The amount of input and output as well as the amount of effect devices varies across the sound cards (Mulder, 2015).

In the previous century, sound processing was only conceivable on high-powered computer systems with high-quality input/output peripherals. With the fast growth of mobile tools, the computational capacity of such tools has improved, and technical characteristics now let audio capture as well as processing straight on mobile tools.

8.3. SOUND RECORDING USING MOBILE DEVICES

Audio data computation patterns are evolving in tandem with the growing use of portable apps in the production chain. Mobile tools are another option for several audio recording and mixing firms to increase their offerings(Shneiderman and Plaisant, 2010). Companies are beginning to provide host apps as an offshoot of their primary software applications, and although they only provide limited capabilities to customers, they are well-suited to touch screens. In certain circumstances, utilizing fingers rather than a mouse to handle the program environment would seem more normal to the users(Prandoni and Vetterli, 2008; Chaffey and Smith, 2013). The capability to change the level and loudness of single audio tracks in a multitrack event, for instance, or the direct manipulation of the FX characteristics of the effects employed, considerably increases the creative part while creating the resultant sound (Austin, 2016).

Of above social media sites handle the bulk of promo video content distribution. The expanding speed, as well as access to mobile internet, as well as access to mobile internet, is driving growth in video media consumption around the globe, not just in the Czech Republic (Speedtest, 2018). Users are connected 24 hours a day, not only because of LTE advancements, and mobile phones are often used as the final distributing module. With their mobile phones and tablets, video customers commonly view promo video items. Nevertheless, the audio track replication could be an issue with some systems. Built-in speakers (various technologies and acoustic characteristics) or integrated headphone outputs and headphones are available to the user (as well as quality and design) (Marinelli, 2009; Ifeachor and Jervis, 2002). Large amounts of video content are available on social media as well as other digital distribution platforms, and the end consumer/user may be overwhelmed. This is why a rising audio mix may have a significant influence on the development (and not only) of promo video slots (Lane et al., 2010).

Mobile technologies have evolved into full-featured recording equipment as a result of advancements in recording technology and apps. The purchasing price is their benefit over the traditional recording equipment (microphone, preamp, AD/DA converter, hard drive). Individual recording elements (of varying technical grades) are merged within one unit in smartphones (Lieb, 2012; Tashev, 2009). Mobile devices may also be utilized as a balanced option to elevated hardware solutions such as microphones, sound cards, and PCs, due to the increasing efficiency of mobile devices. The recording of digital sound is impacted by several factors that work together to generate a cohesive entire. One is the audio resource itself; selecting an appropriate voice, if feasible, provides a significant production benefit. The acoustic recording environment is another factor that influences the recording procedure. It's best to choose an acoustically ideal location here and attempt to minimize any undesired kinds of ambient noise and diffraction. The sampling frequency and bitrate have an impact on the efficiency of the digital record. Certain mobile phones (Apple and Samsung) support a sampling rate of 48Hz / 24bit. The volume of the file is affected by the limit of such factors (Yewdall, 2012).

8.4. TESTING DIFFERENT RECORDING DEVICES

The experiment discussed in this article is limited to the digitizing of audio utilizing various technologies (Lieb, 2012; Chaffey and Smith, 2017). The efficiency of audio input transformation to digital medium is evaluated with the lack of any postproduction changes, since they may have a substantial impact on the final audio output regardless of the input's level. Postproduction stages involve equalization (typically elimination of sub-basses as well as bass suppression along with modification of other equalization regions), dynamic changes (compressors, limiters, and expanders), stereo base correction, and eventual mastering (an increment of volume utilizing multi-band compressors and maximizers) (Koskinen, 2017; Russ, 2012). Additionally, postproduction audio methods may eradicate and significantly enhance the flaws of the recorded version. The most common of such approaches are dynamic signal balance and the elimination of acoustic contaminants via the utilization of suitable processes and plug-ins(Borwick, 2012). Consequently, audio signals captured by smartphones may likewise be optimized via post-production. In latest years, Disney Interactive's user research team has increasingly been requested to do mobile game tests in parallel to the platform and PC-based tests. Disney Interactive currently conducts a large proportion of its user research on mobile games (Hwang and Chen, 2017; Silva et al., 2009). Over the last several years, the user research team at Disney Online has changed to meet the changing demands of the firm. This change in emphasis to the mobile platform prompted more study into updated mobile study recording technologies (Tashev, 2009). Furthermore, mobile games include gestural user interfaces which demand careful consideration. The tiny screen size and touch-based interaction may provide specific design flaws, such as unreadable user interface components, small or overflowing hitboxes, and times in which the player's fingers hide a portion of the display (Vincent et al., 2018; Zölzer, 2008).

It is necessary to examine contacts along with device hardware in parallel to contacts with the game software to evaluate the specific usability problems presented by mobile devices. Due to the form nature of mobile phones, gamers may keep them close to their bodies (Vaseghi, 2008; Smaldino et al., 2008). It tends to make monitoring difficult, particularly for the design teams that must watch from anywhere on the testing site.

The experimental setup comprised a conventional record circuit of an Audio Technica 4047 condenser microphone, a BAE 1073MP preamp, a LUCID 88192 AD/DA converter, and a MacBook Pro computer as the app host (Bevilacqua et al., 2011). Ableton 9 was the recording program we utilized. We employed an Apple iPad tablet (iPad 2017, 32 GB, Wi-Fi cellular), an iPhone cell phone (iPhone 7, 32 GB, black), and a Samsung smartphone as alternate record links to our high-efficiency, single developer (Samsung Galaxy, S8, black). The sound would be transmitted through the Genelec 8130 APM, which is equipped with AES / EBU digital voice (24-bit / 192 kHz) (Hoeg and Lauterbach, 2003).

The capturing of the voice signal is the primary subject of experiments conducted on numerous technical units with distinct factors and implementation performance. As a result, two distinct impulses of two voices, one male as well as the other female, are utilized as the original audio file. The dynamic range of the two produced audio recordings varied, and the speech volume increased with time. Both cited audio recordings showed the various places of the spoken word, ranging from quiet emotions (whispering) to acoustically powerful periods. The recording instrument was set 20 centimeters away from the sound to imitate environmental disruptions. We intended to evaluate the gadgets' capacity to capture audio in natural environments (Whitaker, 2018) (Figure 8.1).

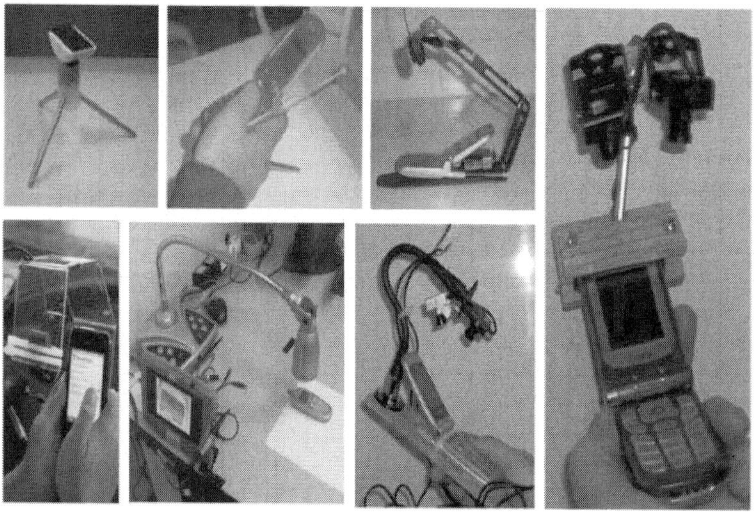

Figure 8.1. (clockwise from top left): last mobile record equipment wireless mobile sled, tripod, phone connected to the tripod, wooden clamping sled, wooden foundation sled, documentation camera over phone, transparent mobile sled.

Source:https://www.wobomart.lk/product/yunteng-yt-228-mini-tripod-for-mobile-phone-and-camera/.

The intensity of the generated signal is influenced by the microphone's input sensitivities and subsequent boosting by the microphone preamplifier at the time of recording. It's sometimes impossible to modify the intake gain and thus impact the sound on smartphones (Watkinson, 2012). Whenever recording loud signals, it also isn't always practical to incorporate the attenuation gain pads (the microphone might overwrite the signal, the digital distortion will occur and the resulting audio signal will be degraded). Changing the distance between the recording equipment and the subject is sometimes the only method to fix these issues. The sound source or the divergence of the recording axis from the microphone insert's directional properties (Tan and Lindberg, 2008) (Figure 8.2).

Figure 8.2. Recording apparatus.

Source: https://www.terramedia.co.uk/media/television/bbc_vera.htm.

The frequency parameters of every track were measured to assess the captured signals both objectively and subjectively (Gold et al., 2011). The subjective assessment contrasted the sound's color alteration to the initial dynamic range maintenance. Individual paragraphs were also examined for clarity. The reference recording was listened to using Neumann KH 310 speakers and AKG and Westone UM Pro 30 in-ear earphones (Fuchs, 2021; Hales et al., 2016). At the time of the reference hearing, it is easy to detect even minor deviations in every one of the tested sequencings thanks to the multitrack features as well as a direct analogy of the actual source data with the report that includes. The subjective assessment was based on a ten-point Likert scale, with 10 being the best quality of outcomes and 0 representing the weakest level of quality (Lee, 1998; Lu et al., 2009) (Table 8.1).

Table 8.1. Subjective Assessment of the Parameters Measured

Recording device	Parametr I. (sound color)	Parametr II. (dynamics)	Parametr III. (intelligibility)
Device I.(AT+BAE+LUCID)	10	10	8
Device II. (Tablet iPad 2017, 32 GB,wifi cellular)	8	7	7
Device III.(iPhone 7, 32 GB, black)	8	6	7
Device IV. (Samsung Galaxy, S8, black)	7	8	7

The frequency characteristics of each recorded session were measured in the second step of the review. In Figure 8.3, graphs of the frequency characteristics of the original and recorded signals are compared. We may conclude from this direct comparison of both signals that each recording device has an impact on the original frequency range (Drotner and Schrøder, 2013). The IV. device was found to have the most significant changes. The majority of recording equipment was unable to deal with the subbase and base band, as shown by the reduction in relevant frequencies. This isn't always a bad thing, since the band below 150 Hz is deleted for the human voice to provide a better grasp of the spoken word (Miluzzo et al., 2008) (Figure 8.4).

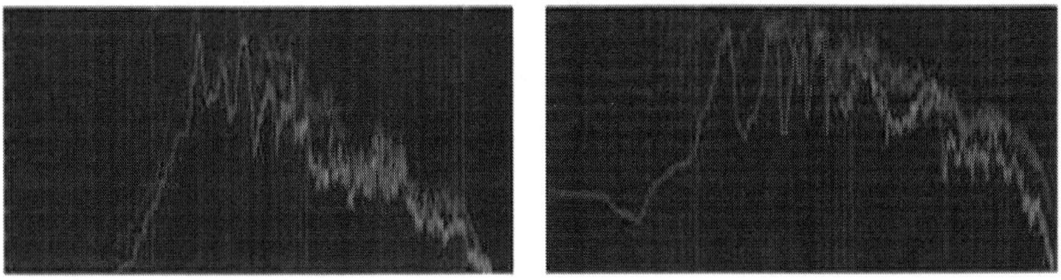

Signal recorded by the Device I

Original signal

Figure 8.3. Recorded signals.

Source:https://www.researchgate.net/figure/The-original-recorded-signal-7_ Fig3_308831525.

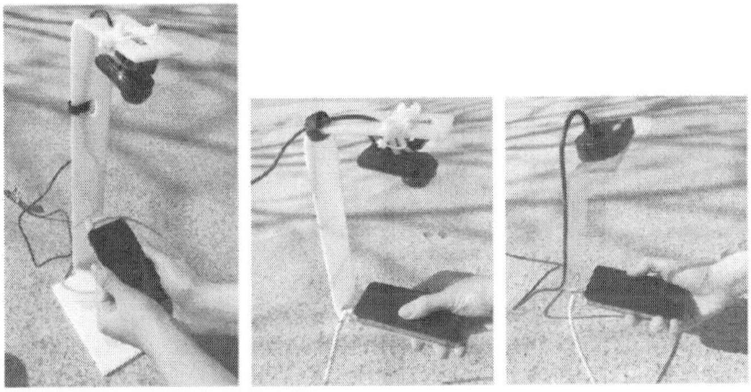

Figure 8.4. The current recording devices, from left to right: White sled, clean sled, above the camera.

Source: https://www.mdpi.com/2077–1312/10/5/555/htm.

8.5. CONCLUSION

The experiment mentioned above convincingly established that mobile devices are capable of recording human voices. Mobile devices may now be utilized for more than simply their core function thanks to modern technology, gadgets, and plugins. The user may therefore receive a highly strong and easy sound recording tool by using the appropriate extensions. We were able to demonstrate a certain degree of competitivity of mobile devices when compared to traditional audio studio equipment via experimental assessments. It's also worth noting that no post-production measures have been taken, which might, of course, considerably improve sound quality if an appropriate technical approach was used. On the other hand, owing to the improper use of post-production instruments, post-production might result in irreparable damage to the sound signal. Other elements (such as the sound environment, sound source, or the recording operator's practical expertise) may also have a substantial impact on the audio recording's quality. Despite these factors that might lead to failures, we feel that mobile devices can be a good alternative to the conventional sound recording chain when it comes to producing material that is acceptable for social media dissemination. However, it's worth noting that social media platforms and the majority of online advertising formats don't always need the finest possible sound quality. Even if mobile devices are a fantastic powerhouse for making typical Facebook or Instagram posts, a conventional studio sound recording chain would be better appropriate for larger advertising initiatives and campaigns.

REFERENCES

1. Aumond, P., Lavandier, C., Ribeiro, C., Boix, E. G., Kambona, K., D'Hondt, E., & Delaitre, P., (2017). A study of the accuracy of mobile technology for measuring urban noise pollution in large scale participatory sensing campaigns. *Applied Acoustics, 117*(1), 219–226.
2. Austin, M. L., (2016). Safe and sound: Using audio to communicate comfort, safety, and familiarity in digital media. In: *Emotions, Technology, and Design* (Vol. 1, pp. 19–35). Academic Press.
3. Bergeron, B., (2005). *Developing Serious Games (Game Development Series)* (Vol. 1, pp. 1–3). Charles River Media, Inc.
4. Bevilacqua, F., Schnell, N., Rasamimanana, N., Zamborlin, B., & Guédy, F., (2011). Online gesture analysis and control of audio processing. In: *Musical Robots and Interactive Multimodal Systems* (Vol. 1, pp. 127–142). Springer, Berlin, Heidelberg.
5. Biswas, T., Mandal, S. B., Saha, D., & Chakrabarti, A., (2017). Coherence based dual microphone speech enhancement technique using FPGA. *Microprocessors and Microsystems, 55*(1), 111–118.
6. Borwick, J., (2012). *Loudspeaker and Headphone Handbook* (Vol. 1, pp. 9–35). CRC Press.
7. Brandstein, M., & Ward, D., (2001). *Microphone Arrays: Signal Processing Techniques and Applications* (Vol. 1, pp. 1–7). Springer Science & Business Media.
8. Chaffey, D., & Smith, P. R., (2013). *eMarketing eXcellence: Planning and Optimizing Your Digital Marketing* (Vol. 1, pp. 1–3). Routledge.
9. Chaffey, D., & Smith, P. R., (2017). *Digital Marketing Excellence: Planning, Optimizing and Integrating Online Marketing* (Vol. 1, pp. 9–16). Routledge.
10. Drotner, K., & Schrøder, K. C., (2013). *Museum Communication and Social Media*. New York: Routledge.
11. Essex, J., & Haxton, K., (2018). Characterizing patterns of engagement of different participants in a public STEM-based analysis project. *International Journal of Science Education, Part B, 8*(2), 178–191.
12. Fuchs, C., (2021). *Social Media: A Critical Introduction* (2nd edn, pp. 19–35). Sage.
13. Fujimoto, K., (2001). *Mobile Antenna Systems Handbook* (4th edn, pp. 3–9). Artech House.

14. Gold, B., Morgan, N., & Ellis, D., (2011). *Speech and audio Signal Processing: Processing and Perception of Speech and Music* (Vol. 1, pp. 5–10). John Wiley & Sons.

15. Hales, S., Turner-McGrievy, G. M., Wilcox, S., Fahim, A., Davis, R. E., Huhns, M., & Valafar, H., (2016). Social networks for improving healthy weight loss behaviors for overweight and obese adults: A randomized clinical trial of the social pounds off digitally (Social POD) mobile app. *International Journal of Medical Informatics, 94*(1), 81–90.

16. Hansen, D., Shneiderman, B., & Smith, M. A., (2010). *Analyzing Social Media Networks with NodeXL: Insights from a Connected World (*Vol. 1, pp. 4–9). Morgan Kaufmann.

17. Hoeg, W., & Lauterbach, T., (2003). *Digital Audio Broadcasting* (Vol. 1, pp. 1–6). New York: Wiley.

18. Hwang, K., & Chen, M., (2017). *Big-Data Analytics for Cloud, IoT and Cognitive Computing* (Vol. 1, pp. 5–10). John Wiley & Sons.

19. Ifeachor, E. C., & Jervis, B. W., (2002). *Digital Signal Processing: A Practical Approach* (Vol. 1, pp. 2–5). Pearson Education.

20. Jones, A. C., & Bennett, R. J., (2015). *The Digital Evolution of Live Music* (Vol. 1, pp. 4–9). Chandos Publishing.

21. Junco, R., (2014). *Engaging Students Through Social Media: Evidence-Based Practices for use in Student Affairs* (2nd edn, pp. 4–6). John Wiley & Sons.

22. Koskinen, I. K., (2017). *Mobile Multimedia in Action* (Vol. 1, pp. 4–8). Routledge.

23. Lane, N. D., Miluzzo, E., Lu, H., Peebles, D., Choudhury, T., & Campbell, A. T., (2010). A survey of mobile phone sensing. *IEEE Communications Magazine, 48*(9), 140–150.

24. Lee, W. C., (1998). *Mobile Communications Engineering: Theory and Applications* (Vol. 1, pp. 9–15). McGraw-Hill Education.

25. Lian, S., (2008). Multimedia content encryption: Techniques and applications. Auerbach Publications, (Vol. 1, pp. 2–15).

26. Lieb, R., (2012). *Content Marketing: Think Like a Publisher--How to Use Content to Market Online and in Social Media* (Vol. 1, pp. 1–5). Que Publishing.

27. Lu, H., Pan, W., Lane, N. D., Choudhury, T., & Campbell, A. T., (2009). SoundSense: Scalable sound sensing for people-centric applications on

mobile phones. In: *Proceedings of the 7th International Conference on Mobile Systems, Applications, and Services* (Vol. 1, pp. 165–178).

28. Manjunath, B. S., Salembier, P., & Sikora, T., (2002). *Introduction to MPEG-7: Multimedia Content Description Interface* (3rd edn, pp. 4–8). John Wiley & Sons.
29. Marinelli, E. E., (2009). *Hyrax: Cloud Computing on Mobile Devices Using MapReduce* (Vol. 1, pp. 9–35). Carnegie-Mellon univ Pittsburgh PA school of computer science.
30. Miluzzo, E., Lane, N. D., Fodor, K., Peterson, R., Lu, H., Musolesi, M., & Campbell, A. T., (2008). Sensing meets mobile social networks: The design, implementation and evaluation of the cenceMe application. In: *Proceedings of the 6th ACM Conference on Embedded Network Sensor Systems* (Vol. 1, pp. 337–350).
31. Pejovic, V., Lathia, N., Mascolo, C., & Musolesi, M., (2016). Mobile-based experience sampling for behavior research. In: *Emotions and Personality in Personalized Services* (Vol. 1, pp. 141–161). Springer, Cham.
32. Prandoni, P., & Vetterli, M., (2008). *Signal Processing for Communications* (Vol. 1, pp. 3–7). EPFL press.
33. Quan-Haase, A., & Sloan, L., (2017). Introduction to the handbook of social media research methods: Goals, challenges and innovations. *The SAGE Handbook of Social Media Research Methods*, *1*(3), 1–9.
34. Russ, M., (2012). *Sound Synthesis and Sampling* (Vol. 1, pp. 9–15). Routledge.
35. Safko, L., (2010). *The Social Media Bible: Tactics, tools, and Strategies for Business Success* (2nd edn, pp. 3–9). John Wiley & Sons.
36. Shneiderman, B., & Plaisant, C., (2010). *Designing the User Interface: Strategies for Effective Human-Computer Interaction* (Vol. 1, pp. 3–6). Pearson Education India.
37. Silva, M. J., Gomes, C. A., Pestana, B., Lopes, J. C., Marcelino, M. J., Gouveia, C., & Fonseca, A., (2009). Adding space and senses to mobile world exploration. Mobile technology for children: *Designing for Interaction and Learning, 1*, 147–169.
38. Smaldino, S. E., Lowther, D. L., Russell, J. D., & Mims, C., (2008). *Instructional Technology and Media for Learning, 1,* 2–8.

39. Speedtest (2018). Speedtest global index. *Ranking Mobile and Fixed Broadband Speeds from Around the World on a Monthly Basis* (Vol. 1, pp. 9–35).
40. Tan, Z. H., & Lindberg, B., (2008). *Automatic Speech Recognition on Mobile Devices and Over Communication Networks* (Vol. 1, pp. 3–10). Springer Science & Business Media.
41. Tashev, I. J., (2009). *Sound Capture and Processing: Practical Approaches* (2nd edn., pp. 3–5). John Wiley & Sons.
42. Turner, J., & Shah, R., (2014). *How to Make Money with Social Media: An Insider's Guide to Using New and Emerging Media to Grow Your Business* (Vol. 1, pp. 9–11). Pearson Education.
43. Van, T. S., (2018). What can mobile do for me?. *The Five Technological Forces Disrupting Security,* (5th edn, pp. 69–80).
44. Vaseghi, S. V., (2008). *Advanced Digital Signal Processing and Noise Reduction*. John Wiley & Sons.
45. Vincent, E., Virtanen, T., & Gannot, S., (2018). *Audio Source Separation and Speech Enhancement* (Vol. 1, pp. 2, 3). John Wiley & Sons.
46. Watkinson, J., (2012). *The Art of Sound Reproduction* (Vol. 1, pp. 4–6). Routledge.
47. Whitaker, J. C., (2018). *The Electronics Handbook* (3rd edn., pp. 1–5). CRC Press.
48. WordStream, (2018). *75 Super-Useful Facebook Statistics for 2018* (4th edn., pp. 1–5).
49. Yewdall, D., (2012). Production sound. *Practical Art of Motion Picture Sound, 1*, 509–530.
50. Zölzer, U., (2008). *Digital Audio Signal Processing, 2*(3), 2–7. John Wiley & Sons.

INDEX